AUTONOMOUS AGENTS

Edited By

George A. Bekey
University of Southern California

Reprinted from
a Special Issue of
AUTONOMOUS ROBOTS
Volume 5, No. 1
March 1998

KLUWER ACADEMIC PUBLISHERS
Boston / Dordrecht / London

AUTONOMOUS ROBOTS

Volume 5, No. 1, March 1998

Special Issue on Autonomous Agents

The cover shows TITAN VIII, a quadruped walking machine developed at Tokyo Institute of Technology in the Laboratory of Professor Shigeo Hirose. Professor Hirose is one of the world's leading designers and builders of autonomous robots.

Distributors for North America:
Kluwer Academic Publishers
101 Philip Drive
Assinippi Park
Norwell, Massachusetts 02061 USA

Distributors for all other countries:
Kluwer Academic Publishers Group
Distribution Centre
Post Office Box 322
3300 AH Dordrecht, THE NETHERLANDS

Library of Congress Cataloging-in-Publication Data

A C.I.P. Catalogue record for this book is available
from the Library of Congress.

Printed on acid-free paper.

Printed in the United States of America

Autonomous Robots, 5, 5 (1998)

Introduction

An agent is a system capable of autonomous actions in a dynamic and unpredictable environment. Hence, agents must be able to perceive the environment, reason with the percepts and then act upon the world. Agents can be purely software systems, in which case their percepts and output "actions" are encoded binary strings. When agents are realized in hardware, and situated in the world, then they are robots.

The First International Conference on Autonomous Agents was held in Santa Monica, California in February 1997. This conference brought together researchers from around the world with interests in agents of many types and varieties. Included were (1) agents inhabiting software environments both in computers and networks ("softbots"); (2) agents which assist a user in the performance of some task ("expert assistants"); (3) synthetic agents which operate in virtual and simulated environments, as in computer animation; and (4) autonomous robots. A number of papers were concerned with basic issues in the study of agents, such as knowledge acquisition, planning, environment modeling, multiple agents and people-agent interaction. Your Editor served as coordinator of the papers on robotic agents. Most of the authors in this group were invited to submit their papers for a second review by this Journal, and this issue presents these revised papers.

The first paper by Fujita and Kitano presents a truly remarkable small autonomous quadruped robot, developed primarily for entertainment. It raises a number of issues concerning robot characteristics which make them attractive to people, as well as the possibility of an open architecture for mobile robots. The second paper considers the perceptual and motor components of a mobile robots to be agents as well, so that the complex behaviors are produced by combinations of lower level agents. The third paper describes a new architecture for autonomous spacecraft control systems, developed as part of NASA's New Millennium program. This architecture will be the first AI system to control an actual spacecraft. The following paper concerns the use of a group of small mobile robots designed to explore and construct maps of an indoor environment, where the host synthesizes partial information provided by each robot into a plausible map. The following paper by Horswill presents a new and powerful agent architecture based on grounding the inferences of an ultra-fast problem solver in continually updated sensor information, using modal logic. One of the traditional problems of intelligent robotics has been the design of architectures for properly relating high-level planning and low-level motor commands. The paper by Zita Haigh and Veloso describes such an architecture which also incorporates learning. The following paper, by Durfee at al, takes this issue further, and describes an architecture which integrates human supervisors, planning agents and robotic agents. The final paper, by Franz et al, presents a vision-based method for learning a topological representation of an open environment by using a graph model.

We consider the field of intelligent agents to be a bridge between artificial intelligence and robotics, and we are pleased to present these papers.

George A. Bekey

Autonomous Robots, 5, 7–18 (1998)

Development of an Autonomous Quadruped Robot for Robot Entertainment *

MASAHIRO FUJITA
D21 Laboratory, Sony Corporation
mfujita@pdp.crl.sony.co.jp

HIROAKI KITANO
Sony Computer Science Laboratory Inc.
kitano@csl.sony.co.jp

Abstract. In this paper, we present **Robot Entertainment** as a new field of the entertainment industry using autonomous robots. For feasibility studies of Robot Entertainment, we have developed an autonomous quadruped robot, named **MUTANT**, as a pet-type robot. It has four legs, each of which has three degree-of-freedom, and a head which also has three degree-of-freedom. Micro camera, stereo microphone, touch sensors, and other sensor systems are coupled with newly developed behavior generation system, which has emotion module as its major components, and generates high complex and interactive behaviors. Agent architecture, real-world recognition technologies, software component technology, and some dedicated devices such as Micro Camera Unit, were developed and tested for this purpose. From the lessons learned from the development of MUTANT, we refined the design concept of MUTANT to derive requirements for a general architecture and a set of interfaces of robot systems for entertainment applications. Through these feasibility studies, we consider entertainment applications a significant target at this moment from both scientific and engineering points of view. [1]

Keywords: autonomous robot, quadruped robot, entertainment, agent architecture

1. Introduction

In this paper, we present **Robot Entertainment** as a new field for using autonomous robots in the entertainment industry. In order to demonstrate the feasibility of Robot Entertainment, we have developed a pet-type legged robot called **MUTANT** (Fig.1), which has four legs and head, each of which has three degrees-of-freedom and is equipped with on-board sensors such as a micro-camera, a stereo microphone, and touch sensors. It is also endowed with a behavior generation system consisting of an instinct/emotion module, a high-level cognition module, and reactive behavior subsystems. MUTANT interacts with people by tonal sounds, and exhibit a large variety of complex behavioral patterns. MUTANT was developed to investigate the feasibility of using robots as an entertainment tool.

We consider entertainment applications an important target at this stage of both scientific and industrial development. The three major reasons are:

Complete Agent: A robot for entertainment requires a completely autonomous physical agent. Instead

*A preliminary version of this paper was presented at the First International Conference on Autonomous Agents in February 1997.

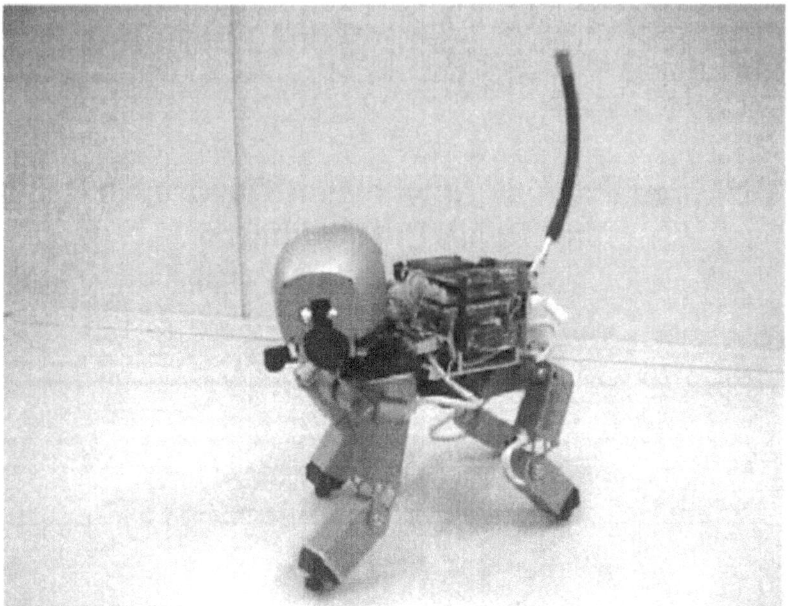

Fig. 1. MUTANT: an fully autonomous pet robot

of focusing on specific perceptual functions such as speech and vision, research on complete agents promotes and accelerates research activities involving the integration of subsystems.

Technology Level: A robot for entertainment can be effectively designed using various state-of-the-art technologies, such as speech recognition and vision, even though these technologies may not be mature enough for applications where they perform a critical function. While there exists special and difficult requirements in entertainment applications themselves, limited capabilities in the speech and vision systems may turn out to be an interesting and attractive feature for appropriately designed entertainment robots.

Emerging Industry: We believe that we will be able to create a completely new market in the near future by introducing this kind of robot products sharply focused on entertainment applications. We strongly believe that after the Gold Rush of the Internet and cyber-space, people will eagerly seek *real* objects to play with and touch. Robot Entertainment provides tangible physical agents and an unquestionable sense of *reality*.

It is expected that various kinds of entertainment applications will be developed for Robot Entertainment systems. The possibility of entertainment using such an autonomous robot is as follows:

Watching Motions of a Robot: Many people enjoy watching motions, gestures, and behaviors of animals in a zoo. Recently, computer controlled dinosaur robots have been very popular in theme parks. We consider that the ability to perform movements of a certain level of complexity is important for robots to really be entertaining.

Interacting with a Robot: We can also interact with a dog by gestures and voices. In addition to watching the motions of the pet animal, the interaction enhances the owner's enjoyment. We consider that interaction is also an important feature of entertainment robot.

Bringing up a Robot: Recently, simulation games, in which players bring up a virtual computer graphics agent, have popular in video game software. In the research field of Artificial Life (A-Life) and Artificial Intelligence (AI), researchers often discuss entertainment applications in this field (Maes, 1995, Kitano, 1994a, Kitano, 1995a).

Controlling a Robot: "Action" games are also popular in video game software. Many people enjoy controlling a car to compete for a time record, or a character to fight with another character controlled by another player. Robot soccer, for example, would be one of the most promising targets of applications. While RoboCup (Kitano, 1997b) focuses of research issues, consumers would certainly be interested in an easier way to play soccer with robots. Thus, controlling a robot will be one of the major application categories in Robot Entertainment.

Creating a Robot: We believe that the creation of a robot is itself entertaining. Recently, many robot contests have opened, where participants had to create robots under some constraints such as weight, cost, and so on. These participants enjoy creating robots implemented with their ideas, and enjoy observing the behavior of the robots they have designed. We believe that not only technically-trained people but also laymen can enjoy creating a robot, if we can provide a friendly development environment.

It is very likely that there will be more possibilities which we cannot imagine at this moment.

To confirm the above possibility of an entertainment robot, we developed an autonomous robot named **MUTANT**. In the following sections, we describe MUTANT, starting from its design concept, followed by the details of its implementation. Then, we discuss the issue of reusability of robot components, in terms of both hardware and software, for developing various kinds of applications, and for creating different robot configurations differing in shape, sensors, actuators, and so on.

2. Design Concept of MUTANT

To confirm the possibility of entertainment using autonomous robot, we developed a pet-type entertainment robot, named MUTANT. First, we defined the design concept of this prototype robot as follows:

Natural Human-Robot Interaction: Considering that interaction is an important feature of entertainment, robots should be capable of interacting with humans in a natural way without any special tools.

Real-World Complete Agent: In addition, robots should be capable of acting in an ordinary room environment, and avoid setting special conditions in terms of sound and illumination.

Complex Behavior: Considering that movements with a certain level of complexity are an important feature of entertainment, robots should be capable of acting with complex movements and complex behaviors.

Reusable Software: Considering that creating a robot is expected to be an important feature of entertainment, robotic software should be reusable in other robotic systems, in order to make easy the creation of other kinds of robots.

Standalone System with Extensibility: Robots should be a standalone pet-type robot, without any cables such as power lines, so that people regard the robot as a real pet. In addition, the robot should be easily extended to remote-operated system, which a user can control as a competition robot.

Our approach to accomplish the above design concept is the following:

- Use of a camera, a stereo-microphone, and a loudspeaker for natural interaction with humans.
- Tonal-language with lip-whistle for robust sound recognition in noisy environments.
- Complex motion with a quadruped mechanical configuration.
- Use of Object-Oriented Programming paradigm and software component technology.

3. Implementation of the Design Concept

3.1. *Human-Robot Interaction*

For a pet-type robot, one key to entertainment is how a robot interacts with the human, and vice versa. We choose to use a camera, a microphone, and a loudspeaker as main sensors and an effector, as shown in Fig.6, in order to allow natural interaction. In addition, we use a three-dimensional acceleration sensor and touch sensors which are used in low-level behaviors.

3.1.1. Micro-Camera-Unit. In order to make a robot small in size and weight and to reduce cost, we developed a Micro-Camera-Unit (MCU) using multi-

Fig. 2. Micro Camera Unit

chip-module technology (Ueda, 1996), as shown in Fig.2. This MCU includes a lens in the same package to obtain a single thin camera module. The size of the MCU is $23 \times 16 \times 4mm$, with pixels 362×492.

3.1.2. DC Geared-Motor. In addition to the MCU, in order to keep the robot small in size and weight, we developed a DC geared-motor for a link module of legs. It consists of a DC motor, five gears, a potentio-meter, and a motor driver chip (Fig.3). A motor case made of magnesium-alloy serves as the outer shell of a leg.

3.2. Real-World Complete Agent

For voice recognition, either Push-to-Talk techniques or a head-set microphone is used in many applications. However, it is better to interact with a robot without any special physical tools, such as a microphone with a switch. To achieve this goal, the following problems has to be resolved:

Noise: In an ordinary room or office environment, there are many noise sources (such as air-conditioning). In addition, a robot itself generates noise when it moves.

Voice Interference: There is also *voice interference* generated by other people in a room. For example, when we demonstrate our robot, people often talk to each other around the robot.

In general, the ability to distinguish the desired sound source in the noisy environment is referred to as the *cocktail party effect*. A human is able to distinguish a target voice from others voices and noise in a cocktail party. Pet-robots also should be able to exhibit a similar capability, possibly with specialized communication methods.

3.2.1. Tonal-language. We believe that in entertainment applications, it is better to use a pitch component, which is robust in ordinary noisy environments. The simplest way of using pitch components for a human to interact with a robot is to use the musical scale, "A-B-C", as the basic components of communication.

This approach can also solve the second problem, voice-interference in a room. In general, the pitch component of a spoken voice in natural conversation does not keep the same frequency, but varies (Fujisaki, 1984). On the other hand, a musical scale signal can keep a constant frequency. Fig.4 shows a time-frequency graph of a mixture of natural conversation and musical tones by a synthesizer, and Fig.5 shows a time-frequency graph of filtered version. We use a "moving-average" filter with a 400msec averaging window.

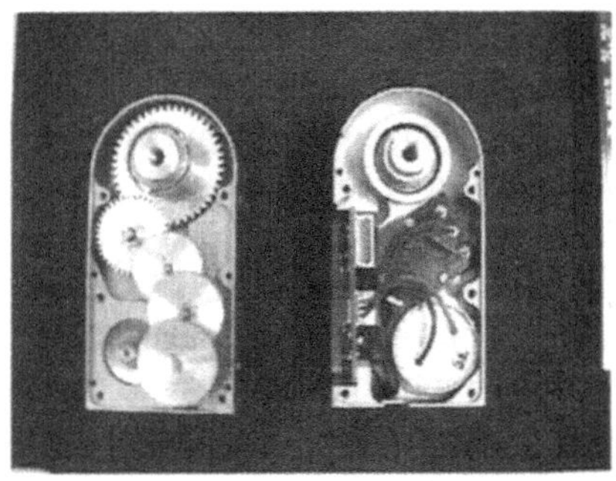

Fig. 3. DC Geared-Motor

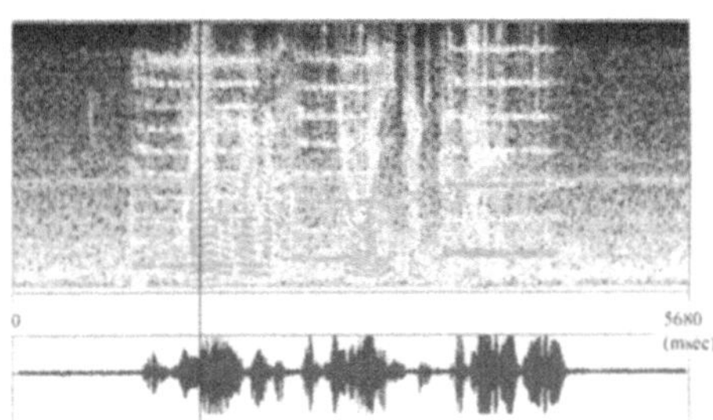

Fig. 4. Time-Frequency Characteristics of Tones with Voices: the sequence of the tones consists of "fa (349.23Hz)", "ra (440.00Hz), and "do (523.25Hz)". While the sequence of the voice is "ohayou gozaimasu. kyou ha ii tenki desune" of a Japanese male speech. At the slice point by the vertical line (the 100th frame), the tone frequency is 349.23Hz.

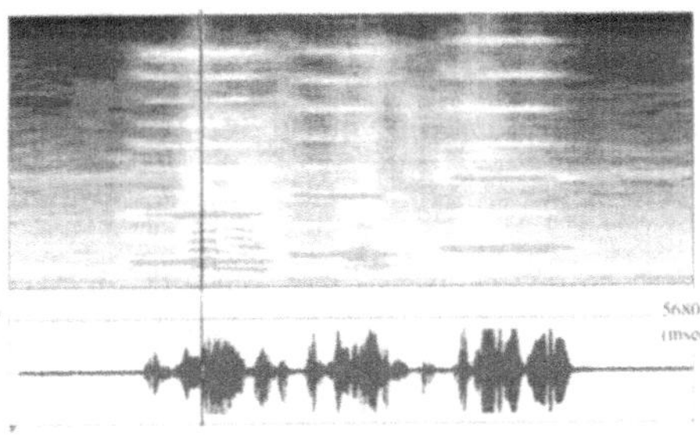

Fig. 5. Filtered Time-Frequency Characteristics of Tones with Voices

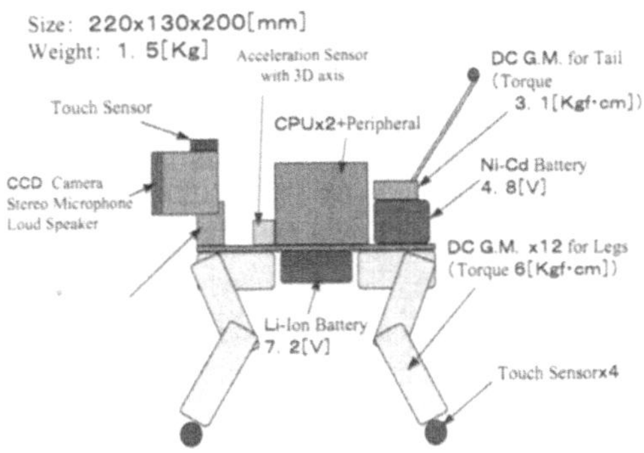

Fig. 6. Mechanical Configuration of MUTANT

Fig.7 shows the spectrums of both the original signal and the filtered signal. In the left-hand graph, the target tone pitch frequency is suppressed by the voice interference; however, in the right-hand graph, the target pitch frequency is recovered by filtering the varying voice pitches.

Since a human can generate such a signal with a lip-whistle or his/her voice, it is possible to interact with a robot without using any special tools in a natural environment.

3.3. *Complex Behavior*

3.3.1. Quadruped Mechanical Configuration. Most of the intelligent autonomous robots are implemented in wheel-based mechanical configurations. Using the Subsumption Architecture, Brooks(Brooks89a) implemented a six-legged walking robot which presented complex behaviors while dynamically interacting with a real-world environment. At least on video, the impact of presenting complex motions of the six-legged robot is bigger than that of wheel-based robot.

We believe that the capability of representation and communication using gesture and motion is very important in entertainment applications. Therefore, the mechanical configuration we chose for MUTANT is that of a quadruped, as shown in Fig.6. The merits of the quadruped configuration are, (1) walking control of a quadruped is easier than that of a biped robot, and (2) when in a posture of sitting, two "hands" are free to move, and allow the display of emotions, or to communicate with a human through hand motions. Thus, because each leg or hand has to be used for various purposes besides walking, we assign three degrees-of-freedom (DoF) for each leg/hand. In addition, we add a tail and three DoF for neck/head so that MUTANT has enough representation and communication capabilities using motions.

3.3.2. Agent Architecture. Another feature in MUTANT for making complex behaviors is to have various kinds of behavior and motion control subsystems, which can at times compete or cooperate. Through competitions and cooperations, complex motions and actions emerge naturally.

There are three dimensions in the competition and cooperation of behavior controls:

Reaction/Deliberation: The question of how to coordinate reactive behavior, as in the subsumption architecture, and deliberation-based behaviors driven by high-level cognitive processing and

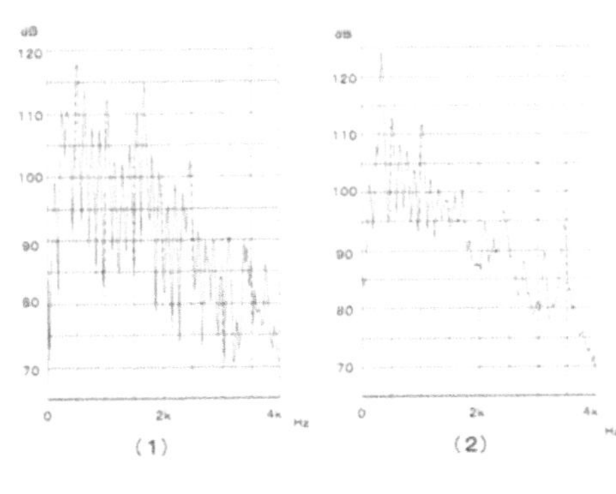

Fig. 7. Spectrum of the 100th frame: (1) Original Data, (2) Filtered Data "showing the recovered tone at 349.23Hz and its harmonics frequency components"

emotional status is one of the central issues in agent architecture. Looking from the aspect of response time from sensor inputs to action outputs, they can be distinguished as (1) reaction with a fast response time, and (2) deliberation with a slow response time. In addition, these two aspects of control are also subject to competition and cooperation as described below.

Instinct/Emotion/High Level Cognition: Within deliberative actions, there are three levels of underlying control subsystems: instinct (or drive), emotion, and high-level cognition. An example of behaviors motivated by instinct (or drive) is an action of moving toward a battery station to *eat* (electricity) when the battery level is low. This class of behavior is triggered, in general, by body needs. An example of behaviors motivated by emotion is the getting angry. Both body needs and situation where the agent is in affect emotional status and their expression. A behavior motivated by high level cognition is, for example, to communicate with a human using motion and voice. This requires some level of planning and reasoning, but is often affected by emotional status.

Hand/Neck/Body: Some of the reactive behaviors can be embedded in each body components. Each part of a robot, such as a right hand, is able to move autonomously with its own control subsystem. For examples, the neck moves to track an object, a hand moves to touch other objects, and the entire robot moves to come close to something. The combination of these motions/actions generated by the parts makes it possible for the robot to exhibit very complex behaviors. In order to generate coordinated reactive behaviors, each body component needs to be cooperative. However, there are cases where two actions are conflicting, and therefore require the selection of one action over the other.

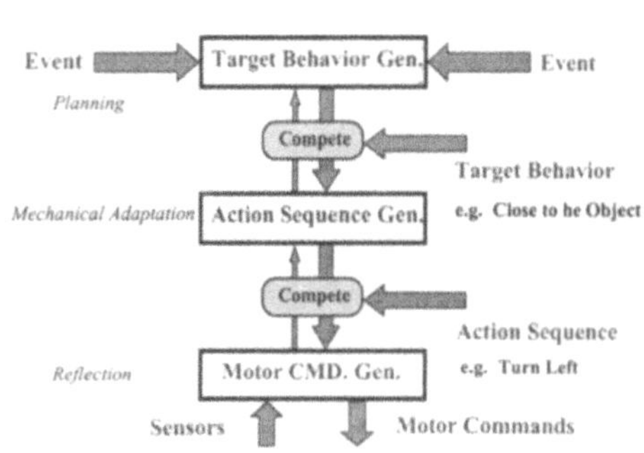

Fig. 8. Layered Architecture for Reaction and Deliberation

These subsystems are integrated by the following methods;

Layered Agent Architecture: We design the agent architecture of MUTANT as a combination of reactive processes and deliberation processes. As shown in Fig. 8, sensory inputs can be processed and a motor command can be sent back immediately by the motor command generator. Higher level processing can be programmed using action sequence generators and target behavior generators.

Instinct/Emotion Module: We designed an instinct/emotion module in the agent architecture so that MUTANT can have a capacity for spontaneous and emotional behaviors. Since this implementation was just a trial for MUTANT to have the capability of complex behaviors, at this moment, we only assign three instinct/emotional states: novelty-boredom, fatigue-activation, and happiness-anger, to MUTANT. Some of the sensors provide input to this module to vary the states which generate some emotional target behaviors such as seeking, sleeping, or getting angry. Although this model is simple, it works to increase the complexity of MUTANT's behaviors. We are currently implementing full-scale emotional/instinct module with rich emotional and body status parameters.

Tree-Structured Architecture: In order to achieve independence of each part of the robot, such as a hand, or a neck, we use the tree structure topology of robot parts configuration, as s hown in Fig.9. In addition, we assign the layered agent architecture to every node in the tree structure. This tree structured agent architecture enables the independence of each part. Competitions and cooperations of these node motivations are carried out by using communication through the tree branches. Currently, for competition and co-operation, MUTANT uses the simplest rule, which is that the control/motivation of the upper part has higher priority. The current implementation of the Tree-Structure has only two nodes, a head and a body.

Fig.10 shows the entire agent architecture of MUTANT, including *Layered Agent Architecture* for a

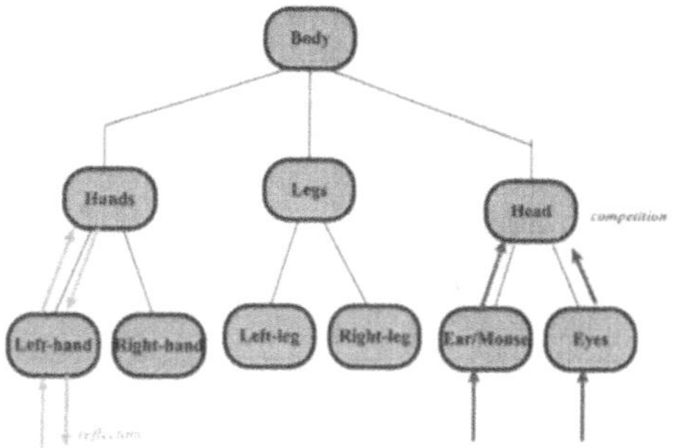

Fig. 9. Tree-Structured Architecture for Sub-System Autonomy

body and a head, *Instinct/Emotion Model*, and *Tree-Structured* Architecture for autonomy of a body and a head.

Fig.11 shows the complex behaviors displayed by MUTANT with its quadruped mechanical configuration and the agent architectures. The "tracking a yellow ball" behavior is carried out reactive behavior, the "karate" behavior is motivated by the instinct/emotional state of anger, the "shake hands" behavior is motivated by color detection of skin color and obstacle distance estimation, and the "sleeping" behavior is motivated by the instinct/emotional state of fatigue.

3.4. *Reusable Software*

One of the major results of recent software engineering is object-oriented programming (OOP) technology (Meyer, 1988), which aims at reusable software production. We use OOP technology for software development for MUTANT.

In addition, we use the Observer Pattern in Design Pattern technology (Gamma, 1995) in order to make each object independent from others. In general, according to the OOP, we define the interface methods for each object, and when using this object, designers for other objects use the defined interface methods. For instance, Object-A processes some data and gets a result, and Object-B uses the interface of Object-A to get this result. In this case, Object-B must know both the interface to get the result and Object-A itself. Thus, Object-B is not independent from Object-A, and this reduces the reusability of Object-B. Observer Pattern solves this problem by defining a subject and observers. In the above case, Object-A is subject to output the data and Object-B is the observer of the data. Each object must know only the type of the data. An observer is dynamically attached to the subject to identify which objects to send the result data to. Thus, the designer of each object does not need to know the other objects. This independence results in software reusability and efficient programming development.

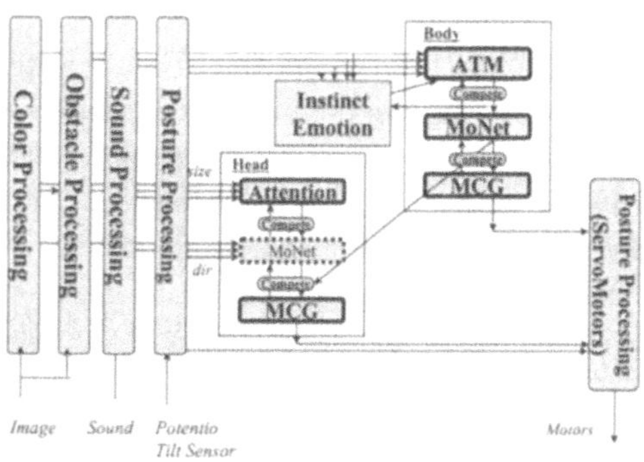

Fig. 10. Entire Agent Architecture of MUTANT

3.5. *Standalone System with Extensibility*

The pet-type robot is a standalone robot, whose battery, CPU board, and every component are on board. Fig. 12 shows another application, with a remote-operated System, for soccer game playing. We add a wireless receiver to MUTANT, and modify the software in the top layer with command interpreter for the wireless command signals.

3.6. *Summary of the Features*

The mechanical and electrical features of MUTANT are summarized in Table.1.

4. Discussion

In order to develop various kinds of Robot Entertainment systems, we believe that reusability of subsystems, in terms of both software and hardware, is very important. This leads us to define **OPENR**[2], a standard architecture and a set of interfaces for Robot Entertainment Systems.

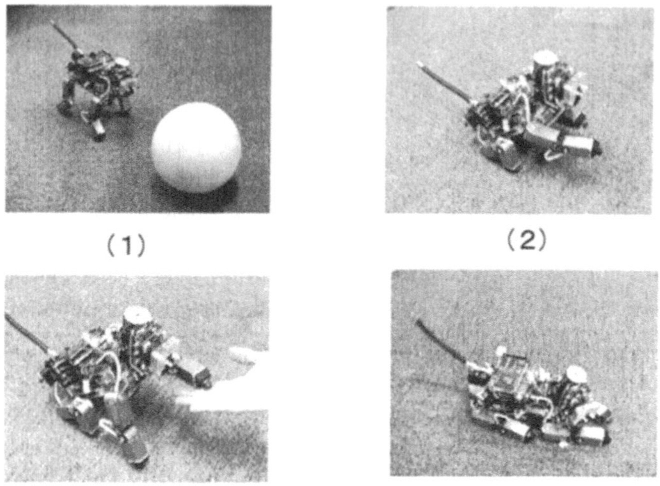

(1) (2) (3) (4)

Fig. 11. MUTANT: (1) tracking a yellow ball, (2) karate action with anger, (3) shake hands, and (4) sleeping

We argue that four types of scalability are essential for such a standard. These are (1) size scalability, (2) category scalability, (3) time scalability, and (4) user expertise scalability.

Size Scalability (Extensibility): Size scalability means that robotic systems can be scaled up in terms of their number of components and number of robots involved in one system. OPENR must be extensible for various system configurations. For example, in a minimum configuration, a robot may only be composed of a few components, and perform a set of behaviors as a complete agent, without any external system. Such a robot can be scaled up in terms of the number of components by adding physical components. It should also be possible to scale up a system by having such robots as sub-systems of large robotic systems. Each robot should be able to work as a part of a larger system, as a remote-operated system.

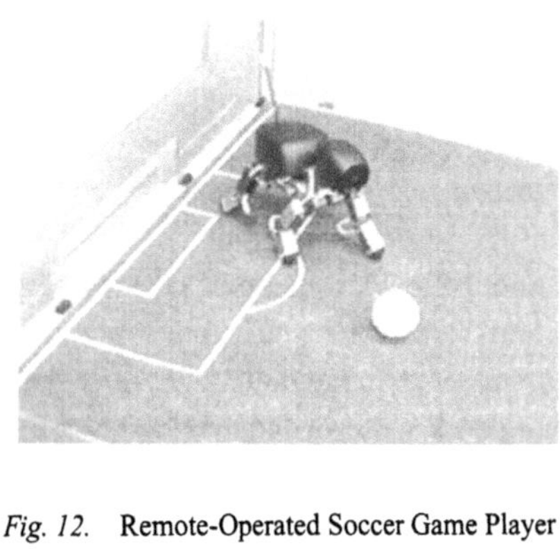

Fig. 12. Remote-Operated Soccer Game Player

Category Scalability (Flexibility): Category scalability ensures that various kinds of robots can be designed based on the OPENR standard. For example, two very different styles of robots, such as a wheel-based robot and a quadruped robot, should be able to be described by the OPENR standard. These robots may have various sensors, such as cameras, infra-red sensors, and touch sensors as well as motor controllers.

Time Scalability (Upgradability): Time scalability guarantees that users can always use up-to-date components, by merely replacing old modules. OPENR must be able to evolve together with the progress of hardware and software technologies. Thus, it must maintain a modular organization so that each component can be replaced with up-to-date modules.

User Expertise Scalability (Friendly Development Environment): OPENR must provide a development environment, both for professional developers and for end-users who do not have technical

Table 1. Features of MUTANT

Size and weight	
Type	quadruped legged robot
Size	$220 \times 130 \times 200[mm]$
Weight	$1.5[Kg]$ (including batteries)
Dof	sixteen
Actuators	
Type	DC geared-motor with a potentio-meter
Torque	$6[Kgf \cdot cm]$ for legs
	$3.1[Kgf \cdot cm]$ for a neck and a tail
Sensors, effectors, and batteries	
Sensors	CCD camera, stereo microphone
	3D acceleration sensor, touch sensors
Effector	a loud-speaker
Battery	Li-ion (7.2V) for electric circuits
	Ni-Cd (4.8) for motor drivers
CPU	
Type	IDT R3052 or R3071 $\times 2$
Clock	about $30[MHz]$
RAM	$8[MBytes]$
Flash ROM	$2[MByes]$

knowledge. End-users may develop or compose their own programs using the development environment. Thus, it should be scalable in terms of the level of expertise that designers of the robot have.

Our approach to meet these requirements are itemized as follows:

- Generic Reference Model, for developing various kinds of robot architectures, with a friendly development environment,
- Configurable Physical Component for reusable robot hardware.
- Object-oriented programming for reusable software components,
- Layering technique for upgradability.

Details of the approach are described in (Fujita and Kageyama, 1997a). We briefly overview OPENR in the Appendix.

5. Related Works

In Robot Entertainment, the shape of each robot should be freely customizable for each application. Most existing robot platforms are not able to customize their shapes, such as YAMABICO(Iida, 1991). Our idea of Configurable Physical Component solves this problem so that designers are able to design various shapes of robots.

Inaba proposed a research and development approach named “Remote-Brain Approach” for autonomous robot (Inaba, 1993). His group has implemented various shapes of robots using this approach, in which computers are not on the robot body, but are remote-hosts with wireless transmitters. Therefore, they can use significant computer power while keeping the size of robot body small. Our architecture can be a standalone robot as well as a remote-controlled robot with wireless extension module using the limited bandwidth of wireless communication channel. The capability of standalone robot without host computers is important when motor commands from host computers cannot be received because of bad wireless channel condition. The humanoid-type robots developed in his laboratory efficiently show the complex motions and behaviors, which we listed as one of the important features for entertainment using autonomous robots.

Robot Entertainment requires satisfy that motions and behaviors of robots be sufficiently complex or unexpected so that people keep an interest in watching or taking care of it. Complex motions and behaviors may be realized using the Subsumption Architecture (Brooks, 1986) or in general Behavior-Based Approach, in which the complexity of motions and behaviors depends on the complexity of environments.

In addition, Robot Entertainment requires that behaviors of the robot should be able to be designed according to a designer's intention. For an autonomous pet-type robot, for example, a designer may want to design a scenario of a robot life, and its personality. For a music lover dancing robot, a designer may want to design its motions. These motions and behaviors are usually expectable and are realized using motion databases or classical AI technology.

The two requirements, complexity and design capability of motions and behaviors, are similar to the requirements, reactive and deliberate behaviors in many Agent Architectures (Hayes-Roth, 1995, Noreils, 1993, Bonasso95). For example, Noreils's architecture (Noreils, 1993) has three layers (levels); functional level, control level and planning level. In the functional model there are many reactive behavior modules which are controlled by the controlled level. In the planning level, deliberate planning is examined. Thus, reactive and deliberate behaviors are both achieved with this architecture.

This technique, the combination of reaction and deliberation, is used in MUTANT as described in the previous section. In our robot, however, planning is more primitive than that of the above Agent Architecture.

Finally, Maes (Maes, 1995) points out that the entertainment area may become much more important for agent research. She introduces ALIVE, a virtual environment allowing wireless full-body interaction between a human and a virtual world which is inhabited by animated autonomous agents. There are some other examples of entertainment agents, such as (Tosa, 1993). Although these agents inhabit virtual worlds, many research topics are very similar to those of Robot Entertainment.

6. Conclusion

We presented Robot Entertainment as a new field of research and industry for real-world AI and robotics. To confirm the possibility of entertainment using autonomous robots, we developed an autonomous robot, named MUTANT, with a design concept that includes (1) natural human-robot interaction, (2) the realization of a real-world complete agent, (3) complex behaviors, (4) software reusability, and (5) extensibility. To realize the design concept, we developed a micro camera unit (MCU) using a multi-chip-module technology, and a DC geared-motor to make MUTANT small in size and weight. The features are summarized in Table.1 (also see Fig.6).

We also tested the tonal-language with musical sound, the agent architecture, and object-oriented programming in the quadruped robot for a pet-type application. Through the development of MUTANT, we are refining our design concept, stressing the importance of the reusability of subsystems or components for establishing Robot Entertainment as an emergent industry. We have discussed a standard architecture and a set of interfaces, and the requirements which such a standard must satisfy.

We believe that Robot Entertainment is significant at this moment, from both the scientific and the engineering points of view, and that we will be able to create a completely new market in the near future.

The authors would like to thank to Dr. Stephane Zrehen at D21 Laboratory, Sony Corporation for his English revising of this paper.

Appendix
Overview of OPENR

A.1. Requirements

As we discussed in secion.4, the four types of scalability are essential for OPENR.

- Size Scalability (Extensibility)
- Category Scalability (Flexibility)
- Time Scalability (Upgradability)
- User Expertise Scalability (Friendly Development Environment)

A.2. Strategy

Our tactics to meet these requirements are itemized as follows:

- Generic Reference Model, for developing various kinds of robot architectures, with friendly development environment,
- Configurable Physical Component for reusable robot hardware.
- Object-oriented programming for reusable software components,
- Layering technique for upgradability.

A.2.1. Generic Reference Model.

We define a generic system functional reference model (GSFRM) composed of Basic System, Extension System and Development System, as shown in Fig.A.1. By defining GSFRM, we are able to construct various kinds of robot systems, such as a standalone robot system, a robot connected with a host computer, a robot with network link, and multi-agent robot system, as in Fig.A.2.

A.2.2. Configurable Physical Component.

OPENR defines common interfaces for all robot components in order to achieve flexible and extensible robot configurations. The physical connection between the robot components is done by a serial bus. For example, using the common components, we can create both a quadruped-type robot and a wheel-based type robot, as shown in Fig.A.3.

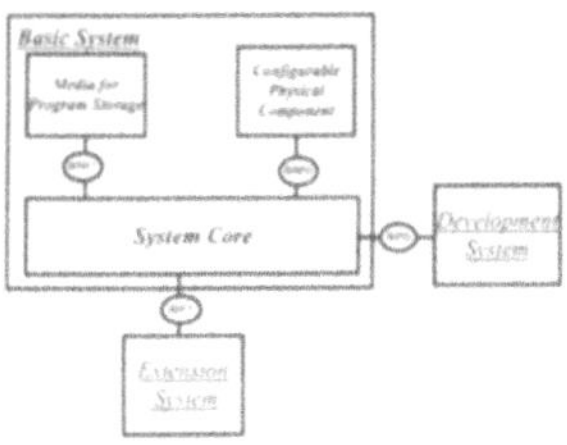

Fig. A.1. Generic System Functional Reference Model and Basic System Functional Reference Model

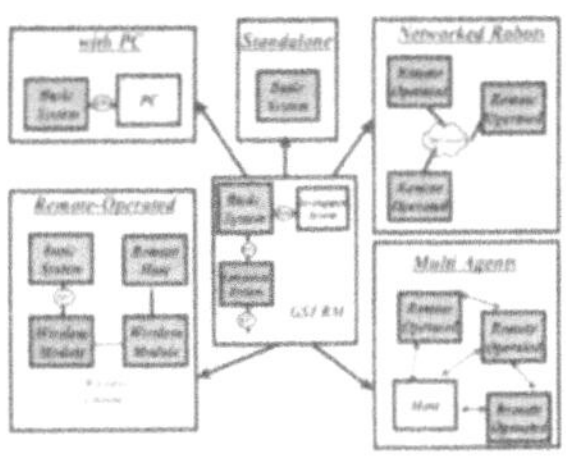

Fig. A.2. Various Architectures derived from GSFRM

Fig. A.3. Various styles of robot with CPC

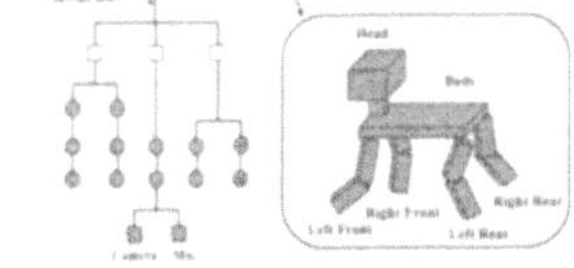

Fig. A.4. Configurable Physical Component

In addition, every CPC has non-volatile memory with (1) functional properties, such as an actuator and a camera, and (2) physical properties such as the three dimensional size of a part.

With this information, it is possible for the host controller of the serial bus to know what kinds of components are part of the robot. Furthermore, as shown in Fig.A.4. in managing the tree connected structure, the host controller can get information on how all the components are connected to each other. This mechanism makes it possible for users to create their own robot configurations with favorite components.

A.2.3. Software Component.

The idea of decomposing the system into components is extended to the software architecture of a robot. The key idea is to employ an object-oriented operating system, Aperios
(Yokote, 1992), and to keep each software object as independent as possible. The Observer Pattern technique is useful for this purpose as we verified in the MUTANT software architecture.

A.2.4. Layering.

The purpose of layering is flexibility and easy development of hardware and software components. OPENR divides each functional element into three layers, Hardware Abstraction Layer (HAL), System Service Layer (SSL), and Application Layer (APL).

Hardware abstraction software in the HAL makes it possible for users to reuse device driver software by providing a standard interface defined by OPENR. The SSL provides basic services to users. The Virtual Robot described above is one such example. The APL provides a more convenient service than that of the SSL. The Designed Robot is one of such example.

Notes

1. Some parts of the contents in this paper are the same as in the paper in the 1st international conference on Autonomous Agents 1997(Fujita and Kageyama, 1997a). In this paper, we emphasize **Robot Entertainment** itself rather than the **OPENR** which was the main issue in Autonomous Agents97. We emphasize a design concept of a prototype quadruped robot, and we describe how we realize the concept, including agent architecture issue.
2. **OPENR** is a trade mark of Sony Corporation

References

Albus, J. S. 1996. The Engineering of Mind. In *Proceedings of the fourth international conference on simulation od adaptive behavior (SAB96, From animals to animats 4)*, pp.23–32.

Bonasso, R. P., Kortenkamp, D., and Miller, D. 1995. Experiences with an Architecture for Intelligent, Reactive Agents. In *IJCAI Workshop on Agent Theories, Architectures, and Languages*.

Brooks, R. A.. 1986. A Robust Layered COntrol System for a Mobile Robot. *IEEE Journal of Robotics and Automation*, RA-2(1):14–23.

Brooks, R. A. 1989. A Robot that Walks: Emergent Behavior from Carefully Evolved Network. *Neural Computation*, 1:253–262.

Firby, R. J. 1994. Task Networks for Controlling Continuous Processes. In *Proceedings on the Second International Conference on AI Planing Systems*.

Fujisaki, H. and HIrose, K. 1984. Analysis of Voice Funcdamental frequency contours for declarative sentenses of Japanease. *J. Acoust. Soc. Jpn.*, 5(4):233-242.

Fujita, M. and Kageyama, K. 1997. An Open Architecture for Robot Entertainment. In *Proceedings of the First International Conference on Autonomous Agents*, Marina del Ray, pp.234–242.

Fujita, M., Kitano, H. and Kageyama, K. 1997. A Legged Robot for RoboCup based on OPENR. In *The First International Workshop on RoboCup in conjunction with IJCAI-97*, Nagoya, pp.69–74.

Gamma, E., Helm, R., Johnson, R., and Vlissides, J. 1995. *Design Pattern: Reusable Object-Oriented Software*, Addison-Wesley.

Hayes-Roth, B. 1995. An architecture for adaptive intelligent systems. *Artificial Intelligence*, 72(1-2):329–365.

Lida, S. and Yuta, S. 1991. Vechicle command system and trajectory control for autonomous mobile robots. In *Proceedings of the IEEE/RSJ International Workshop on Intelligent Robots and Systems 91*, pp.212–217.

Inaba, M. 1993. Remote-Brained Robotics: Interfacing AI with Real World Behaviors. In *Proceedings of the 6th International Symposium on Robotics Research (ISRR6)*, pp.335–344.

Kitano, H. (ed). 1994. *Proceedings of AAAI Workshop on Entertainment and AI/Alife*, Seattle.

Kitano, H. (ed). 1995. *Proceedings of AAAI Workshop on Entertainment and AI/Alife*, Montreal.

Kitano, H., Asada, M., Kuniyoshi, Y., Noda, I., and Osawa, E. 1997. RoboCup: The Robot World Cup Initiative. In *Proceedings of the First International Conference on Autonomous Agents*, Marina del Rey, pp.340–347.

Kitano, H., et al. 1997. RoboCup: A Challenge Problem for AI. *AI Magazine*, Spring:73–85.

Noreils, F. R. 1993. Toward a Robot Architecture Integrating Cooperation between Mobile Robots. Application to Indoor Environment. *The International Journal of Robotics Reseach*, 12(1):79–98.

Maes, P. 1995. Artificial Life meets Entertainment: Lifelike Autonomus Agents. *Communication of the ACM: Special Issue on New Horizons of Commercial and Industrial AI*, pp.108–114.

Meyer, B. 1988. *Object-Oriented Software Construction*, Prentice Hall.

Soukup, J. 1994. *TAMING C++: Pattern Classes and Persistence for Large Projects*, Addison-Wesley.

Tosa, N. 1993. Neurobaby. In *SIGGRAPH-93 Visual Proceedings, Tomorrow's Realities*, ACM, pp.212–213.

Ueda, K. and Takagi, Y. 1996. Development of Micro Camera Module. In *Proceedings of the 6th Sony Research Forum*, pp.114–119.

Yokote, Y. 1992. The Apertos Reflective Operating System: The Concept and Its Implementation. In *Proceeding of the 1992 International Conference of Object-Oriented Programing, System, Languages, and Applications*.

Masahiro Fujita is a senior reseach scientist at D21 Laborotary, Sony Corporation. He received a B.A. degree in Electoronics and Communication from the Waseda University, Tokyo, in 1981, and a M.S. degree in Electrical Engineering from the University of California, Irvine, in 1989. His research interests include Modular Robotics, Neural Networks, Visual and Sound Early-Perception, Face Recognition, Attentional Mechanism, and Emotional Model.

Hiroaki Kitano is a senior researcher at Sony Computer Science Laboratory. He received a B.A. in physics from the International Christian University, Tokyo, and a Ph.D. in computer science from Kyoto University. He was a visiting researcher at Carnegie Mellon University during 1989-1993. Kitano received the Computers and Thought Award from the International Joint Conference on Artificial Intelligence in 1993. Kitano is a founder of RoboCup, and the Chair of The RoboCup Federaion. His reseach interests include RoboCup, computational molecular biology, engineering use of the mophogenesis process, and evolutionary systems.

Autonomous Robots 5, 19–28 (1998)

Basic Visual and Motor Agents for Increasingly Complex Behavior Generation on a Mobile Robot

MARIA C. GARCIA-ALEGRE AND FELICIDAD RECIO
Instituto de Automática Industrial (IAI), Consejo Superior de Investigaciones Cientificas (CSIC), 28500 Arganda del Rey, Madrid, Spain
maria@iai.csic.es

Abstract. Present work addresses the guidelines that have been followed to construct basic behavioral agents for visually guided navigation within the framework of a hierarchical architecture. Visual and motor interactions are described within this generic framework that allows for an incremental development of behavior from an initial basis set. Basic locomotion agents as, Stop&Backward, Avoid, and Forward are implemented by means of fuzzy knowledge bases to deal with the uncertainty and imprecision inherent to real systems and environments. Basic visual agents as, Saccadic, Find_Contour, and Center are raised under a space-variant representation pursuing an anthropomorphic approach. We illustrate how a complex behavior results from the combination of lower level agents always connected to the basic motor agents. The proposed methodology is validated on a caterpillar mobile robot in navigation tasks directed by an object description.

Keywords: mobile robots, agents, architectures, foveated vision, locomotion, fuzzy behaviors, behaviors composition

1. Introduction

For years Artificial Intelligence researchers have devoted most of their efforts to complex problem solving models disregarding real world applications. Emphasis has been placed on declarative knowledge, rational selection, and problem solving strategies (Simon, 1969; Nillson, 1971; Newell, 1990). In contrast to these traditional approaches, reactive and behavior-based systems propose a different methodology to design structures with low level of deliberative complexity created to function in dynamic, uncertain and noisy environments (Brooks, 1986, 1991; Maes, 1990).

In recent decades, there has been increasing interest and attention focused on the design and construction of models that account for relevant interactions with real environments and their significance in the proposed solution. Developed architectures are engaged in assembling knowledge, perception and action, within distributed systems able to cope with limited knowledge and resources in rarely structured worlds (Albus, 1992; Maes, 1991).

To account for major topics, hybrid architectures have emerged as a way to overcome the lack of prediction inherent to highly reflex behaviors, and the absence of real world constraints of algorithmic and heuristic models of Artificial Intelligence (Arkin, 1990; Garcia-Alegre et al., 1993; Crowley, 1994; Hayes-Roth, 1995; Van den Velde, 1995).

The more widespread models for synthesis of artificial behavior are those based upon the schema and agent concepts. Schema theory (Arbib, 1975), was initially drawn up to deal with complex systems in real domains, as a bridge among cognitive science, brain theory, and distributed artificial intelligence, providing a distributed model of computation.

On the other side, Agents theory evolves from Minsky (1986) proposition of the "Society of Mind" wherein members played a role quite similar to that of

*A preliminary version of this paper was presented at the First International Conference on Autonomous Agents in February 1997.

schemas. Initially translated to the robotics world by Brooks (1986) to build up robot controllers, it has been applied in behavior generation for situated robots.

Present work proposes a generic framework to incrementally build up a Multi-Agent hierarchy for ever complex behaviors generation. The model starts from an initial set of basic vision and locomotion agents, with the ultimate objective of deriving robots behaviors for fetch_deliver and guide_visits tasks. Basic agents are analyzed and compared to those designed in group behavior generation. Finally a case study is thoroughly analyzed for a situated robot that aims to follow a person in our laboratory.

2. Concepts and Definitions

Hybrid architectures are used in a wide range of fields, from Artificial Intelligence, Robotics and Software Engineering to Neuroscience, Brain Theory, Ethology and Psychology. This gave rise to a lack of consensus in the meaning of the terms more frequently used as: agent, agency, schema and behavior.

Schema is defined (Arbib and Cobas, 1991; Corbacho and Arbib, 1995) as a modular entity involving data structure and control. It has been applied in robots' navigation as a model for multiple concurrent motion control processes (Arkin, 1989). Some of the relevant characteristics of the schemas theory, useful for our comparative purpose are:

- Network topology close to the structural elements. Disregard a top_level executor since schema instances can combine their effects by cooperative or competitive distributed processes.
- Construct of knowledge representations through assemblage of schemas, for relevant objects and actions involved. Permit different implementations, either biological or technological.

An agent is understood as a process capable of perception, computation and action, that may or may not be physical; and behavior is defined as a control law addressed to effectively achieve and maintain a goal (Mataric, 1994).

Herein are defined the concepts that are been used from now on:

Agent: Formal unit of architecture which holds for both deliberative and reactive abilities without any constraint on its complexity (Garcia-Alegre et al., 1995). In this work agent and behavior will be indistinctly used in the sense of a computational and control process addressed to reach or to maintain a goal. The agent notion has been purposively established as general as possible, as what matters is the observable behavior not the nature of the mechanism by which it is achieved (McFarland, 1993).

Basic Agents: Elementary building blocks needed to solve a specific class of problems (i.e., collision-free motion). Agents have been created with a modular criteria in mind, in the sense of being able to generate other agents by combining basic ones. This main feature implies a scalable, flexible and incremental model. Scalable to grow through heterogeneous processes and processors to avoid running out of resources, flexible to elude "a priori" constraints on the size and complexity of each software module and incremental to allow the addition and deletion of agents.

Level: Corresponds to a set of agents that share a common language with their interlocutors at adjacent levels. The language is composed of a set of terms related to perception, knowledge representation, and actions. A level gravitates around a specific representation of the system and environment and may have a short or long term memory. This spatio-temporal window record may support future learning capabilities to improve the observable behavior competing each level (i.e., local piloting).

A notion similar to that of basic agents is being used for group behavior generation in a fleet of robots by Mataric, namely "basis behaviors" (Mataric, 1994, 1995). In that work some fundamental group behaviors, dependent on both the domain and the goal, are selected as a minimal set with appropriate composition operators. Such behaviors are decomposed in terms of individual structural or functional low level agents as: AVOID, FORWARD, TURN, MOVE_TO, namely HOMING for the group.

Basic agents, here selected to agree with both the definition and the composition rules for those "basis behaviors", being all basis and basic gross grain behaviors.

In respect to the level concept, present interpretation do not accord to Minsky suggestion on the term agency as in current organization of agents, intra-level interaction do not occur (Minsky, 1986, 1994). Our level definition somehow resembles the layer of competence of the subsumption architecture (Brooks, 1986), but present notion proposes some additional characteristics as explicit knowledge representation and adaptation/learning (Guinea et al., 1993).

3. Selection of Basic Locomotion and Vision Agents

A great variety of motion and vision related tasks can be described and achieved in terms of dynamic sequencing of some basic agents. The set of lower level agents that have been selected are called basic in the sense that they form one of the minimal set for generating other agents of ever increasing complexity.

The proposed set of basic agents is not the unique nor the one with minimum components, but it is the one that allows for the completion of currently proposed tasks and of those expected in a near future (i.e., PICK _ UP _ BOOKS _ IN _ JAMES _ DESK, MOVE _ TO _ ENTRANCE _ DOOR).

3.1. Locomotion Agents

Low level motion agents have been extensively studied during last years and they usually correspond to basic reflex behaviors of approaching, running away and non-colliding. These agents are survival instinctive behaviors present in all mobile creatures. Present selection is based on the belief that complex behavior, result from the combination, direct summation or temporal switching, of basic agents through simple arbitration rules.

Three Basic Agents are selected at the locomotive domain, namely: STOP&BACKWARD, AVOID and FORWARD. They are proposed in a decreasing degree of emergency. All three are embedded in a common skin: the piloting level, sharing a local world representation centered in the mobile robot. This representation is a mapping of ultrasonic, infrared, and tactile readings to distance and orientation of the obstacles in its vicinity.

STOP&BACKWARD: Driven by the tactile and infrared sensors, is needed to recover from unexpected events due to sensor failures or very acute obstacle shape. It appears as a branching in the actions, for identical perceptual conditions. As an example: 1) IF close-to-contact conditions THEN FORWARD, or 2) IF close-to-contact conditions THEN STOP&BACKWARD, where first rule produces a push behavior and second one a safe navigation. The appearance of more than one agent, introduces the concept of upper level to arbitrate the activation (action selection) of lower level agents.

AVOID: Triggered by ultrasonic sensors. Its objective is the detection of obstacles. It is in charge of the supervision of medium range distance all around the robot. When a safety distance is reached a detour behavior is displayed; the selection of this value modulates the mobile behavior from a very precautious one to a high speed one.

FORWARD: Driven also by the ultrasonic sensors. It aims the detection of free-space. Moves the robot to a location specified in polar coordinates and is provided with time-out mechanisms to avoid sticking to a behavior under sensors/actions failures or knowledge inconsistencies.

These agents are modeled as fuzzy knowledge bases for decision making on the robot heading and speed. They also take into account proprioceptive sensors readings at a rate of one hundred and eighty milliseconds. External and proprioceptive sensory variables are represented as fuzzy sets, and are defined through trapezoidal membership functions (Gasós et al., 1992).

Table 1. Fuzzy knowledge basis of AVOID agent for DELTAv control.

R1: **IF** dist_obst is FAR. AND. speed is LOW
THEN DELTA v is POSITIVE
R2: **IF** dist_obst is NEAR. AND. speed is HIGH
THEN DELTAv is NEGATIVE
R3: **IF** dist_obst is FAR. AND. speed is HIGH
THEN DELTAv is ZERO
R4: **IF** dist_obst is NEAR. AND. speed is LOW
THEN DELTAv is ZERO

The set of fuzzy rules that structure the knowledge base of the AVOID agent, is shown in Table 1. These rules are designed keeping in mind simplicity.

Key system-environment states are selected and defined as the relevant sensory patterns that trigger a basic agent; any other pattern not considered, generates a default action: stop. In current implementation basic agents are triggered by mutually exclusive system-environment perceptual states, so as to circumvent explicit intra-level arbitration.

The finest grain level decomposition of motion behaviors, for individual mobile creatures, is performed in terms of the agents: FORWARD, BACKWARD, TURN _ RIGHT, TURN _ LEFT, that obviously are connected with the lower control structures.

3.2. Visual Agents

They are designed to selectively react to visual stimuli in a dynamic environment under the requirements of a specific task (Tistarelli and Sandini, 1992). Human vision forms multi-resolutional images at the visual cortex due to the non-uniform photoreceptors distribution. Biologically inspired, the logarithmic-polar transformation has proved to be one of the most suitable approaches (Panerai et al., 1995) to implement this non-homogeneous sensing for space selective artificial vision.

To provide a space variant vision a foveated sensor has been implemented, mapping the image Cartesian plane to an image cortical plane. This transformation provides variable resolution in the visual field with a high resolution small inner area, namely fovea, that gradually decreases towards the periphery. This log-polar sensor highly reduces processing resources but requires more activity to accurately explore new image regions enlightening the classical paradigm of memory versus activity.

When looking for a specified target in a scene, visual agents must gather and coordinate all the information available about the target. This is the commitment of the SEARCH&TRACK agent, inspired by the ocular movements in human vision, that arbitrates sequential combinations of the basic agents. These processes according to our initial definitions are agents or behaviors in spite of being only perceptual or motor, very much like the perceptual and motor fine grain schemas of Arbib theory.

Three agents are selected as basic visual behaviors, taking into account that in the robot the eye and the head are rigidly coupled together (Bustos et al., 1995):

SACADDIC: Shifts the focus of attention toward different scene locations, by selecting targets from the periphery of the visual field. Targets are all kind of objects attributes according to the higher level descriptions. Outputs are short and rapid motions within the field of view. With such a design, attention is focused only in very small image regions.

FIND_CONTOUR: Extracts a closed contour from the edge filtered image. This contour is adjusted by considering the inner and outer gradient directions, and the relative position of the contour adjacent pixels. Once stabilized, inner region is classified in accordance with the target attributes: gray level and size.

CENTER: Maintains the recognized contour in the fovea region by tracking the contour centroid at the inner sensor area, at a rate of 25 fps. Periphery is dynamically defined as the outer region of the current segmented contour.

The real-time dynamic interaction of these Agents is arbitrated by higher level Agents as: SEARCH&TRACK or SEARCH&COUNT, and allows for the solution of a multiplicity of cases that concern visual stimuli guided navigation (Recio, 1995).

Eye movements and retinal segmentation, intentionally have been kept at the lowest vision level as three processes associated to short and rapid camera movements, while smooth long term camera displacements having a more purposive component are grouped at the next vision level. All three agents jointly generate an unique eye observable behavior in the sense of tightly coupling perception and action at a high activation ratio.

4. A Hierarchy of Agents

A hierarchy of levels is proposed to handle complexity, wherein each level is associated with a degree of abstraction, time-response, and system-environment representation. This hierarchical approach is deeply grounded in the ethologic behavioral model of a hierarchy of functional and structural behaviors, namely centers, proposed by Tinbergen (1951).

The key questions underlying this loosely coupled architecture are:

- The design of a formalism that both deals with and do not drastically separate, motor control mechanisms from purely cognitive strategies (Mataric, 1997).
- To ease the incremental process of new agents construction, by reusing all the functionality of previously generated, starting from the set of basic agents.
- To reduce development time, permitting several users to concurrently program different modules.

Levels in the hierarchy communicate through bidirectional channels, that have no associated protocol; consequently it must be defined between communicating agents based upon perception and acting possibilities.

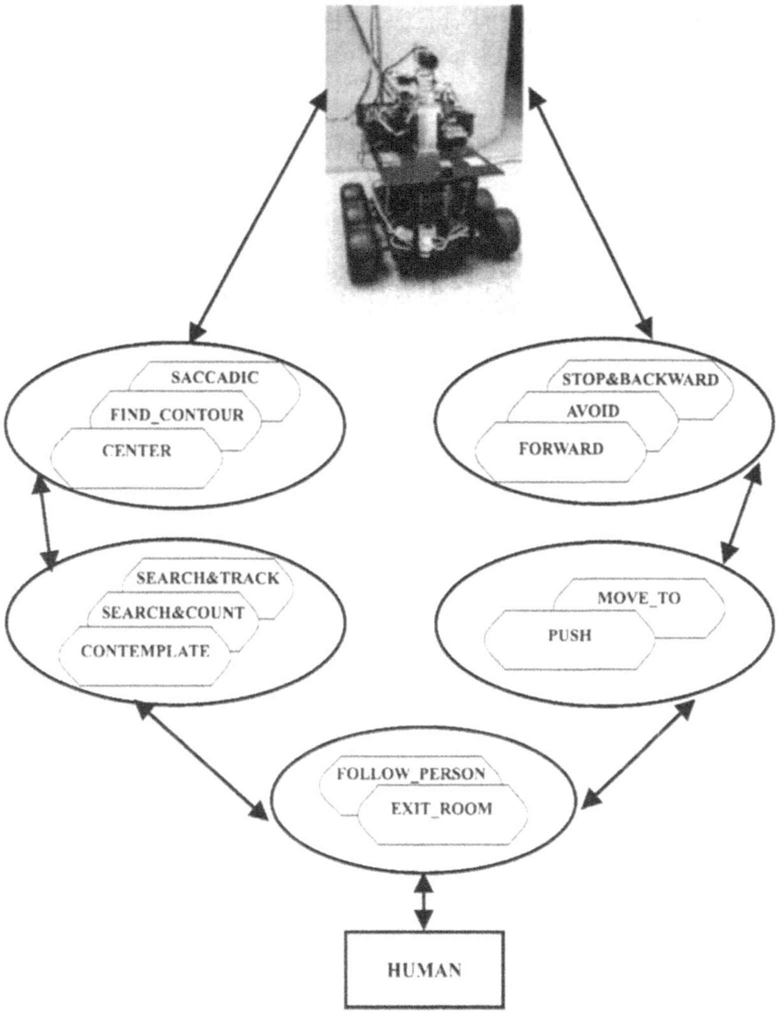

Figure 1. Multi-agents hierarchical architecture.

From a global perception-action perspective, there are two information flows: goal-driven which progress top-down, and event-driven which flows bottom-up. The first one depends on the arbitration mechanism to use lower level abilities of the system. The second one relies on the upwards information propagation of the robot-environment state.

Arbitration: Mechanism by which a basic agent allows the instantiation of its own behavior. Arbitration is a run-time process that depends on the system goal and situated robot. Agents of a level are arbitrated by agents of higher levels that requests some expression of their competence in order to reach their goal. In practice, there is no guarantee that the request will be achieved, hence time-out mechanisms and alternative strategies must be suggested to avoid system deadlocks.

Propagation: Perceptual information which flows from lower to higher levels. An agent propagates information when it is explicitly requested to do so or when an arbitrated demand can not be accomplished within the required constraints.

This general framework has been kept as simple as possible under the assumption that complexity must grow in parallel for all three: robot, environment and tasks. The developed architecture is schematically shown in Fig. 1.

The agents architecture displays a hierarchical topology, with a top-level executor human being and gross grain behavioral agents, that in current implementation do not hold intra-level interactions. In contrast, schema model underlies a network topology which could easily approach fine grain structures, and do not need top-level firing due to concurrent schema interactions.

The architecture for group behavior emergence exhibit a more open topology, highly dependent upon system, goals and domains. Group behavior is always gross grain, and up to date do not embed large cognitive tasks.

5. Visual and Motor Interactions: Approaching a Visual Stimulus

Interactions between basic locomotion and vision agents has been demonstrated with the caterpillar type robot: BOSS-IAI which has been built at the IAI/CSIC. The robot is endowed with a rotating ultrasonic sensor, one degree of freedom (dof) and a vision system. This system incorporates a conventional CCD camera (512 × 512 pixels) on a two dof Pan/Tilt unit. All sensors are located at the body, a six wheels platform with two degrees of freedom. The hardware architecture has evolved from a eight processors Transputer-based on a single Pentium/90 to the current Dual Processor/PPro 180 MHz, Linux O.S., that communicates with host Unix machines via Ethernet. Unix processes are connected by sockets, and communication protocols allow each level to behave as both a sockets server and a client, accepting and requesting connections at run time. The system assumes motion decoupling between the eye/head and the body (Guinea et al., 1995).

The missions for this robot are all addressed to solve a particular class of problems: those worried about a "safe navigation guided by visual stimuli" at indoor office domains. The success in reaching the goals repertoire, implies the accomplishment of tasks with ever increasing level of detail.

Activation sequencing will interactively depend on the dynamic of the system and environment information flow. Flow propagates bottom-up as perception/generalization mechanisms and top-down as planning alternatives/decision_making mechanisms.

We now present and analyze an instance of the class of problems already mentioned, for a situated system. The "FOLLOW-PERSON" agent, with some specified attributes, is activated by a human being. Next SEARCH&TRACK and MOVE_TO agents are triggered and these in one's turn arbitrate the activation sequence of the basic agents under the effect of real world interactions. A top-down cognitive model of what the robot has to see is used to control vision and to drive the sequences of rapid eye movements and structure fixation required for an efficient human-like run over the image (Noton and Stark, 1971).

<_>**FOLLOW-PERSON** [Target Attributes]
[WHAT?(gray level, size, free-space, safety-distance),
WHERE?(pan and tilt area, proximity-distance)
HOW?(slow, fast, accurate, fuzzy for camera/head and body)]

__>**FOLLOW-PERSON** [Actions]<=>

<_>**SEARCH&TRACK** [Target Attributes]
[WHAT?(gray level, size),
WHERE?(pan and tilt area),
HOW? (slow, fast, accurate, fuzzy)]

||

<_>**MOVE** [Target Attributes]
[WHAT? (free-space),
WHERE? (J=camera / head - body°, r = proximity-distance),
HOW? (slow, fast, accurate, fuzzy)]

AND

__>**SEARCH&TRACK** [Actions]<=>
While active goal & not time-out
if **FIND_CONTOUR** not recognized then
SACCADIC to next target direction
else **CENTER** target

||

__>**MOVE** [Actions]<=>
While free-space distance>proximity distance
FORWARD
if free-space distance=proximity distance
STOP&BACKWARD
else **AVOID** or **STOP&BACKWARD**

__>REPORTS (to **HUMAN**)

The SEARCH&TRACK agent will take into account target attributes before determining the actions sequence to be tentatively triggered at a first glance, notifying the FOLLOW-PERSON agent on success or

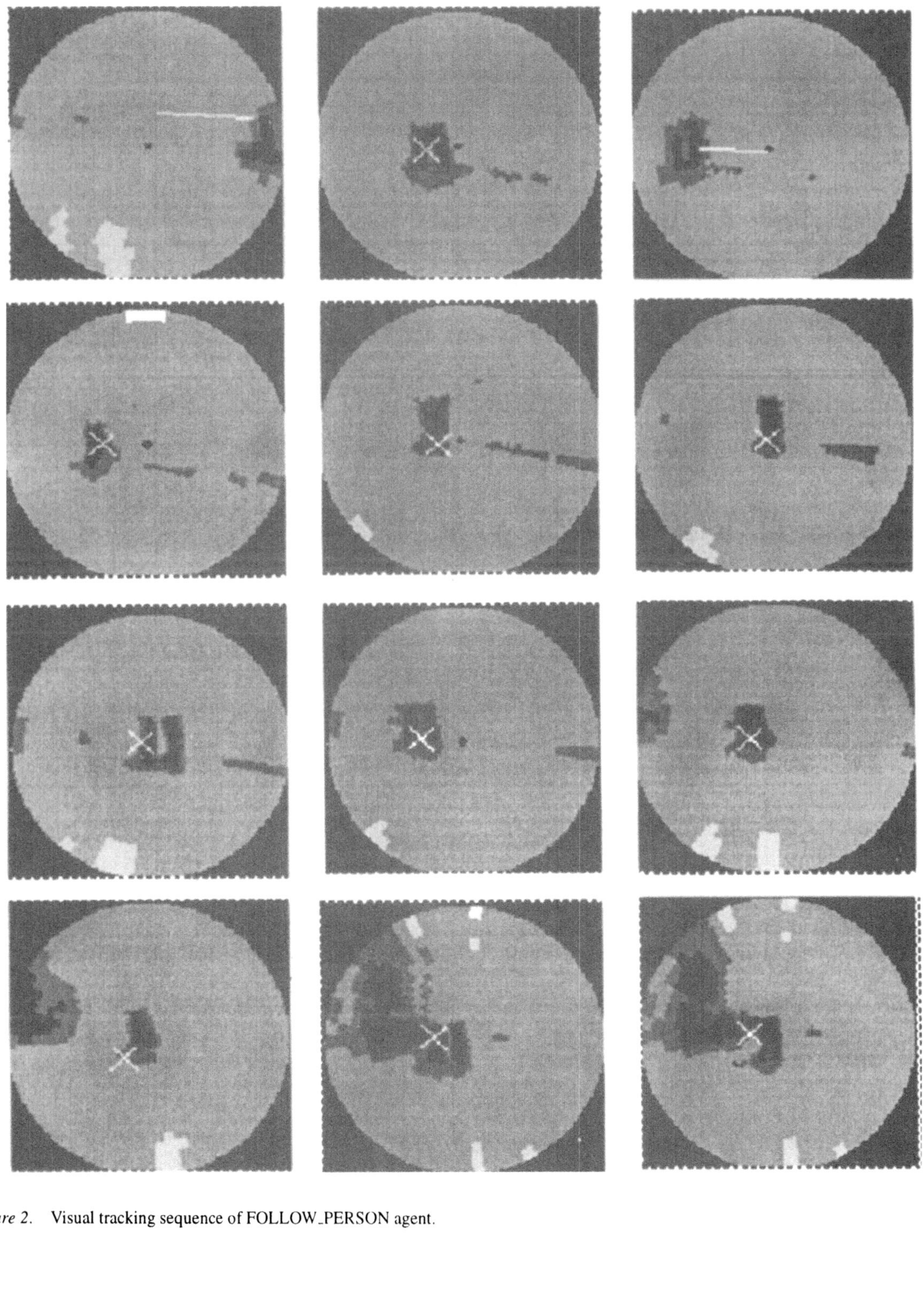

Figure 2. Visual tracking sequence of FOLLOW_PERSON agent.

failure. This agent arbitrates three agents at the adjacent lower level, see Fig. 1. When a target is lost, a search in periphery is activated in the target emergence direction. Here again the interaction of the three agents along with their eventual arbitration, generates an emergent behavior able to cope with a new and ever more complex class of problems.

FOLLOW-PERSON agent relies on the SEARCH& TRACK agent ability to recover from tracking failures by continuously alternating between searching in the scene area and smoothly pursuing the centroid of the recognized targets.

In very extreme situations both agents will report from failures to FOLLOW-PERSON, that will either proceed to look for a new strategy or send some message to the human being. MOVE agent ensures mobile robot safe motion avoiding obstacles within the direction and distance constraints, solving most unexpected events. MOVE agent modulates the ultrasonic sensors readings to obtain a local representation of the surrounding free-space.

The PUSH agent has been generated as an alternative to the MOVE agent to derive a different action regardless of having identical system-environment perception. It is defined as:

PUSH [Target Attributes]
[WHAT? (obstacle)
WHERE?(J=obstacle-direction, r= target_d, push_d),
HOW? (slow, fast, accurate, fuzzy)]

PUSH [Actions]<=>
if free-space distance<= safet_d .AND. curren_d < (target_d - safety_d)
THEN **AVOID** or **STOP&BACKWARD**
else **FORWARD**

Herein the first premise is concerned with the control under specific target and safety distances. A tracking sequence of a human being wearing black trousers, is shown in Fig. 2 as an example of the FOLLOW-PERSON behavior.

Unexpected observable behaviors sometimes emerge as an effect of the interaction among the physical system-environment, the proposed goal, and the explicit arbitration algorithm embedded at each behavioral agent. Experimental results over a wide range of initial conditions and scenarios, point to coordination as the main cue for well-suited observable behavior. This explicit coordination of both experimental and expert knowledge can be greatly improved by using some learning techniques as proved by Mataric in the optimization of group behavior (Mataric, 1996).

6. Discussion

A set of basic behaviors for visually guided tasks in an office-like domain, is selected, implemented and demonstrated with a physical robot in a dynamic environment. It aims at an interactive tasks repertoire accomplishment focused on the development of ever more friendly and ease interactions with humans.

Such basic behaviors are embedded in a general framework that reuse their abilities to derive ever more complex and abstract agents to incrementally reach the set of tasks required for a specific application.

The design characteristics of the proposed architecture allowed four scientist to develop in parallel vision and locomotion related agents according only to the pre-defined communication protocols.

The set of basic agents for locomotion tasks are formulated by means of fuzzy rules bases approaching human expert strategies for mobile piloting. They generate smooth trajectories to attain a location within some tolerances. Environmental and structural knowledge is injected to the systems as arbitration strategies and target attributes values, to help pruning the perceptual space and get real time performance.

The proposed architecture is an appropriate framework for agent-environment interaction descriptions, that allows for an easy incremental spreading of the class of problems that can be tackled with the available perception and action system.

Acknowledgments

Present work is funded in completion by Spanish CICYT Grants: TAP94-0171 and TIC96- 1392- C02-01. Authors want to thank D. Guinea, Angela Ribeiro and P. Bustos for fruitful discussions; G. Sánchez for valuable help in sensors integration and Sara Steele for final proof-reading. Thanks also to Prof. G.A. Bekey and anonymous referee for valuable comments and suggestions that have led to a much improved paper.

References

Albus, J.S. 1992. RSC: A reference model architecture for intelligent control. *IEEE Computer, Special Issue on Computer Architectures for Intelligent Machines*.

Arbib, M.A. 1981. Perceptual structures and distributed motor control. *Handbook of Physiology—The Nervous System II. Motor Control*, Amer Physiol. Soc.

Arbib, M.A. and Cobas, A. 1991. Schemas for prey-catching in frog and toad. *From Animals to Animats: Proc. of the First Int. Conf. on Simulation of Adaptive Behavior*, pp. 142–151.

Arkin, R.C. 1989. Motor schema-based mobile robot navigation. *Int. J. Robotics Res.*, 8(4):92–112.

Arkin, R.C. 1990. Integrating behavioral, perceptual, and world knowledge in reactive navigation. *Robotics and Autonomous Systems*, 6:105–122.

Brooks, R.A. 1986. A robust layered control system for a mobile robot. *IEEE Journal of Robotics and Automation*, RA-2:14–23.

Brooks, R.A. 1991. Intelligence without representation. *Artificial Intelligence*, 47:139–160.

Bustos, P., Recio, F., Guinea, D., and Garcia-Alegre, M.C. 1995. Cortical representations in active vision on a network of transputers. In *Proc. of First ECPD Int. Conf. on Advanced Robotics and Intelligent Automation*, Athens, pp. 259–264.

Connell, J.H. 1991. SSS: A hybrid architecture applied to robot navigation. In *IEEE Int. Conf. on Robotics and Automation*, Nice, pp. 2719–2724.

Corbacho, F. and Arbib, M.A. 1995. Learning to detour. *Adaptive Behavior*, 3(4):419–468.

Garcia-Alegre, M.C., Ribeiro, A., Gasós, J., and Salido, J. 1993. Optimization of fuzzy behavior-based robot navigation in partially known industrial environments. In *The Third Int. Conf. on Industrial Fuzzy Control and Intelligent Systems*, Houston, TX, pp. 50–54.

Garcia-Alegre, M.C., Bustos, P., and Guinea, D. 1995. Complex behavior generation on autonomous robots: A case study. In *Proc. of IEEE on System, Man, and Cybernetics*, Vancouver, B.C., pp. 1729–1734.

Gasós, J., Garcia-Alegre, M.C., and Garcia, R. 1992. Fuzzy strategies for the navigation of autonomous mobile robots. *Fuzzy Engineering towards Human Friendly Systems*, IOS Press: Amsterdam, Holland.

Guinea, D., García-Alegre, M.C., Kalata, P., Lacaza, A., and Meystel, A. 1993. Robot learning to walk: An architectural problem for intelligent controllers. *Eighth IEEE Int. Symp. on Intelligent Control*, Chicago, IL, pp. 493–498.

Guinea, D., Sánchez, G., Bustos, P., and García-Alegre, M.C. 1995. A distributed architecture for active perception in autonomous robots. *IEEE Int. Conf. Syst. Man and Cyber.*, Vancouver, Canada, pp. 1740–1745.

Hayes-Roth, B. 1995. An architecture for adaptive intelligent systems. *Artificial Intelligence, Special Issue on Agents and Interactivity*, 72(1–2):329–365.

Luck, M. and d'Inverno, M. 1995. A formal framework for agency and autonomy. In *Proc. First Int. Conf. on Multi-Agent Systems*, San Francisco, CA, pp. 254–260.

Maes, P. 1990. Situated agents can have goals. *Designing Autonomous Agents: Theory and Practice from Biology to Engineering and Back*, MIT Press/Bradford Books: Cambridge, CA.

Maes, P. 1991. A bottom-up mechanism for behavior-selection in an artificial creature. In *From Animals to Animats, Proc. of the First Int. Conf. on the Simulation of Adaptive Behavior*, MIT Press/Bradford Books: Cambridge, MA.

Mataric, M. 1994. Interaction and intelligent behavior. M.I.T. Artif. Intell. Lab., Massachusetts Inst. Technol., Cambridge, MA, A.I. Memo. 1495.

Mataric, M. 1995. Issues and approaches in the design of collective autonomous agents. *Robotics and Autonomous Systems*, 16(2–4): 321–331.

Mataric, M. 1996. Learning in multi-robot systems. *Lectures Notes in Artificial Intelligence (LNAI)*, 1042:152–163.

Mataric, M. 1997. Studying the role of embodiment in cognition. *Cyber. & Systems*, 28(6):457–570.

McFarland and Bösser, T. 1993. *Intelligent Behavior in Animal and Robots*, MIT Press: Cambridge, MA.

Minsky, M.L. 1986. *The Society of Mind*, Simon & Schuster Ed.: New York.

Minsky, M.L. 1994. A conversation about agents. *Communication of the ACM*, 22–29.

Newell, A. 1990. *Unified Theories of Cognition*, Harvard University Press: Cambridge, MA.

Nillson, N.J. 1971. *Problem Solving Methods in Artificial Intelligence*, McGraw Hill.

Noton, D. and Stark, L. 1971. Eye movements and visual perception. *Scientific American*, 224(6):34–43.

Panerai, F., Capurro, C., and Sandini, G. 1995. Space variant vision for an active camera mount. LIRA-TR 1/95. LIRA-DIST, University of Genoa.

Recio, F. 1995. Tracking monocular. Informe Técnico. 09/95, Dept. of Systems, IAI/Consejo Superior de Investigaciones Científicas.

Simon, H. 1969. *The Sciences of the Artificial*, MIT Press: Cambridge, MA.

Steels, L. 1994. Building agents with autonomous behavior systems. In *The Artificial life route to Artificial Intelligence: Building Situated Embodied Agents*, Lawrence Erlbaum Assoc.: New Haven, MA.

Tinbergen, N. 1951. *The Study of Instinct*, Oxford University Press.

Tistarelli, M. and Sandini, G. 1992. Dynamic aspects in active vision. *CVGPI: Image understanding*, 1(56):108–129.

Van den Velde, W. 1995. Cognitive architectures-from knowledge level to structural coupling. In *The Biology and Technology of Intelligent Autonomous Agents*, Springer Verlag.

Maria C. Garcia-Alegre is a staff scientist in the Systems Department at the Instituto de Automática Industrial (IAI), Consejo Superior de Investigaciones Científicas (CSIC) in Madrid since 1990. She received a Ph.D. degree in Physics at the University of La Laguna, Tenerife in 1983, where she held a permanent position as an Assistant Professor. She was a visiting Professor at the Applied Machine Intelligence & Robotics Laboratory, at Drexel University, Philadelphia in 1992 doing research on hierarchical architectures in control systems. Participant and leader of several European and National Research Projects related to Mobile Autonomous Robots, Fuzzy Logic, Distributed Architectures and Artificial Vision Systems. She is a member of several professional societies, IEEE, FLAT, AEIA. Her work has been published in professional journals, books and conference proceedings in Computer Science, Fuzzy Logic and Robotics. She has also served as proposal referee for the Spanish National Science Foundation (CICYT).

Felicidad Recio is a Ph.D. candidate in the Systems Department at the Instituto de Automática Industrial (IAI), Consejo Superior de Investigaciones Científicas (CSIC) in Madrid. She received the B.Sc. in the University Complutense, Madrid. She is interested in artificial vision systems for robotics and in active vision mechanisms.

Autonomous Robots 5, 29–52 (1998)

An Autonomous Spacecraft Agent Prototype

BARNEY PELL
Caelum Research Corporation, NASA Ames Research Center, MS 269/2, Moffett Field, CA 94035
pell@ptolemy.arc.nasa.gov

DOUGLAS E. BERNARD, STEVE A. CHIEN, AND ERANN GAT
Jet Propulsion Laboratory, California Institute of Technology, 4800 Oak Grove Drive, Pasadena, CA 91109
douglas.e.bernard@jpl.nasa.gov
chien@aig.jpl.nasa.gov
gat@aig.jpl.nasa.gov

NICOLA MUSCETTOLA AND P. PANDURANG NAYAK
Recom Technologies, NASA Ames Research Center, MS 269/2, Moffett Field, CA 94035
mus@ptolemy.arc.nasa.gov
nayak@ptolemy.arc.nasa.gov

MICHAEL D. WAGNER
Fourth Planet Inc., 220 Main Street, Suite 204, Los Altos, CA 94022
mwagner@4thplanet.com

BRIAN C. WILLIAMS
NASA Ames Research Center, MS 269/2, Moffett Field, CA 94035
williams@ptolemy.arc.nasa.gov

Abstract. This paper describes the New Millennium Remote Agent (NMRA) architecture for autonomous spacecraft control systems. The architecture supports challenging requirements of the autonomous spacecraft domain not usually addressed in mobile robot architectures, including highly reliable autonomous operations over extended time periods in the presence of tight resource constraints, hard deadlines, limited observability, and concurrent activity. A hybrid architecture, NMRA integrates traditional real-time monitoring and control with heterogeneous components for constraint-based planning and scheduling, robust multi-threaded execution, and model-based diagnosis and reconfiguration. Novel features of this integrated architecture include support for robust closed-loop generation and execution of concurrent temporal plans and a hybrid procedural/deductive executive.

We implemented a prototype autonomous spacecraft agent within the architecture and successfully demonstrated the prototype in the context of a challenging autonomous mission scenario on a simulated spacecraft. As a result of this success, the integrated architecture has been selected to fly as an autonomy experiment on Deep Space One (DS-1), the first flight of NASA's New Millennium Program (NMP), which will launch in 1998. It will be the first AI system to autonomously control an actual spacecraft.

Keywords: autonomous robots, agent architectures, action selection and planning, diagnosis, integration and coordination of multiple activities, fault protection, operations, real-time systems, modeling

*A preliminary version of this paper was presented at the First International Conference on Autonomous Agents in February 1997.

1. Introduction

The future of space exploration calls for establishing a "virtual presence" in space. This will be achieved with a large number of smart, cheap spacecraft conducting missions as ambitious as robotic rovers, balloons for extended atmospheric explorations and robotic submarines. Several new technologies need to be demonstrated to reach this goal, and one of the most crucial is on-board spacecraft autonomy.

In the traditional approach to spacecraft operations humans on the ground carry out a large number of functions including planning activities, sequencing spacecraft actions, tracking the spacecraft's internal hardware state, ensuring correct functioning, recovering in cases of failure, and subsequently working around faulty subsystems. This approach will not remain viable due to (a) round trip light time communication delays which make joysticking a deep space mission impossible and (b) a desire to limit the operations team and deep-space communications costs.

In the new model of operations, the scientists will communicate high-level science goals directly to the spacecraft. The spacecraft will then perform its own science planning and scheduling, translate those schedules into sequences, verify that they will not damage the spacecraft, and ultimately execute them without routine human intervention. In the case of error recovery, the spacecraft will have to understand the impact of the error on its previously planned sequence and then reschedule in light of the new information and potentially degraded capabilities.

To help bridge the gap between the old operations model and the new one, we needed to learn about the spacecraft domain and requirements, develop an approach to the problem, and demonstrate to both the AI community and the spacecraft community that our approach was viable for the problems the spacecraft community actually encounters. To this end, we teamed up with some of the best spacecraft engineers to develop and demonstrate an architecture integrating AI tools with traditional spacecraft control. The challenge was to demonstrate complete autonomous operations in a very challenging context: simulated insertion of a realistic spacecraft into orbit around Saturn. The mission scenario included trading off science and engineering goals and achieving the mission in the face of any single point of hardware failure.[1] This Saturn Orbit Insertion (SOI) scenario was proposed by experienced spacecraft engineers who had participated in several previous planetary missions. Although simplified,[2] it still contains the most important constraints and sources of complexities of a real mission, making it the most difficult challenge in the context of the most complicated mission phase of the most advanced spacecraft to date (Pell et al., 1996). Furthermore, the demonstration had to be accomplished in the very short time frame of about six months, at which point the management of NASA's New Millennium Program (NMP) was to decide on the technology plan for the Program's first technology demonstration mission.

As we addressed the task, we found a number of properties that made this domain challenging and interesting from an architectural perspective. First, a spacecraft must be able to carry on autonomous operations for extended time periods in the presence of tight resource constraints and hard deadlines. Second, spacecraft operation requires high reliability in spite of limited observability of the spacecraft's state. Third, spacecraft operation is characterized by concurrent activity across a large number of different subsystems. Hence, all reasoning methods need to reflect the spacecraft's concurrent nature. These properties challenge the capabilities of most current architectures for autonomous robotics, which emerged to meet the requirements of the mobile robots domain. For example, very few mobile robot architectures support concurrent temporal planning or diagnosis in the presence of possibly faulty sensors, much less the integration of these two styles of reasoning which are necessary in the autonomous spacecraft domain.

The unique requirements of this domain led us to the New Millennium Remote Agent (NMRA) architecture. The architecture integrates traditional real-time monitoring and control with (a) constraint-based planning and scheduling, to ensure achievement of long-term mission objectives and to effectively manage allocation of scarce system resources; (b) robust multi-threaded execution, to reliably execute planned sequences under conditions of uncertainty, to rapidly respond to unexpected events such as component failures, and to manage concurrent real-time activities; and (c) model-based diagnosis, to confirm successful plan execution and to infer the health of all system components based on inherently limited sensor information.

NMRA is a hybrid architecture. Each of the heterogeneous components is a state-of-the-art, general-purpose system that had been applied to solving specific subtasks in the domain. Since the components had the capability to support the domain requirements individually, the major challenge was in their integration. Novel features of the integration include support for robust closed-loop generation and execution of concurrent temporal plans and a hybrid procedural/deductive executive.

After six months of effort, the NMRA architecture was successfully demonstrated on the simulated SOI scenario. The scenario turned out to be among the most complex handled by each of the component technologies and furthermore placed strong constraints on how the components could be integrated. This success resulted in the inclusion of NMRA as an autonomy experiment in the first NMP mission, Deep Space 1 (DS-1), which is scheduled to launch in mid-1998. This will be the first AI system to autonomously control an actual spacecraft.

In this paper, we report on the implemented architecture and describe the characteristics of the spacecraft domain which posed constraints on the architecture and its implementation. The rest of the paper is organized as follows. Section 2 contains a description of the spacecraft domain and the Cassini SOI scenario. Section 3 highlights the features of the domain which drove our architectural decisions and compares the spacecraft domain to the mobile robotics domain. Section 4 provides an overview of the architecture, paying careful attention to how the components and their integration addressed these domain requirements. It also discusses each of the major elements of the architecture in some detail. Section 5 provides details on the implementation and discusses the magnitude of the modeling and implementation effort. Section 6 compares our architecture with related work. Section 7 concludes the paper and discusses important areas for future work.

2. Scenario

2.1. *Introduction*

The Cassini SOI was used as the scenario for developing and testing the NMRA prototype. This scenario was chosen by spacecraft engineers at JPL, because it represents one of the most challenging and well-studied problems in spacecraft operations. It entails maneuvering a complex spacecraft (Cassini) with multiply-redundant systems into orbit around Saturn, while capturing science imagery of both the rings and the planet itself and down-loading science and engineering data to the Earth. The scenario centers around the mission-critical Main Engine burn, which slows the spacecraft to the proper velocity for achieving Saturn orbit. Any error in the start time, duration, or vector of the burn will result in mission failure. Consequently, redundant spacecraft systems (e.g., switches, gyros, and even a backup Main Engine) must be pre-configured and ready in case of any failures.

A simplified version of Cassini was used for modeling the prototype spacecraft, and the SOI scenario was condensed into a set of goals and constraints. An example sequence satisfying the goals and constraints was also provided by the spacecraft engineers for reference. The challenge to the autonomous system was not to duplicate this sequence, but rather to plan and execute tasks in such a manner that all constraints were satisfied. Finally, a set of guidelines was established for running the scenario and handling simulated failures.

2.2. *Guidelines*

The following guidelines were established for the scenario:

1. Achieve the mission goals even in the event of any single point hardware failure.
2. Consider the SOI burn a special event that, for robustness, requires that all critical subsystems operate in their highest reliability modes.
3. Although multiple independent simultaneous failures are not considered credible, multiple sequential failures, spaced far enough apart to allow recovery of one before considering the next, are considered credible and must be accommodated.

2.3. *Goals*

The following goals define the SOI scenario:

- Use the main engine to insert the spacecraft into Saturn orbit.
- Acquire and return science images of Saturn during approach.
- Acquire and return science images of Saturn's rings near closest approach.
- Assure that the camera is protected from ring particles during ring-plane crossing.

2.4. *Constraints*

The models of the spacecraft as understood by the planner form the context for achieving the above goals. These models constrain the choices that the planner may make, force certain tasks to be ordered, and force the addition of tasks to allow the goals to be achieved. For the SOI scenario, the following constraints significantly affect the resulting plan:

- Available spacecraft electrical power is limited; each operating mode of each assembly requires a predefined power allocation.
- Available science data storage is limited; there is not enough room to accommodate both the Saturn approach and Saturn ring images simultaneously
- Only one spacecraft pointing direction may be commanded at a time. This couples the science imaging activity, the orbit change activity, the Earth communication activity, and the ring safety activity since all require some spacecraft axis to be pointed in a particular direction (e.g., antenna toward earth).
- A main engine burn requires several preparatory steps prior to engine ignition.

Appendix A provides additional details about the SOI scenario.

3. Domain and Requirements

The spacecraft domain places a number of requirements on the software architecture that differentiates it from domains considered by other researchers. In this section, we discuss the requirements of the spacecraft domain and contrast the domain with the mobile robots (mobots) domain which has been the focus of much of the work in autonomous robotics.

3.1. *Spacecraft Domain*

There are three major properties of the domain that drove the architecture design.

First, a spacecraft must be able to carry on *autonomous operations for long periods of time* with no human interaction. This requirement stems from limitations of deep-space communication and the desire to cut operating expenses. As an example of deep-space communication limitations, the Cassini spacecraft is blocked from Earth communications for a period of about 10 hours during SOI because Saturn is between the spacecraft and earth. Hence, Cassini must perform the critical orbit insertion activities (including fault responses) without any opportunity for human intervention. Similarly, a spacecraft investigating another planet may be on the "dark side" of the planet for a period of weeks, months, or years, during which time it must operate completely autonomously. Even in cases where communications would be physically possible, the costs of communicating (using relay satellites and ground stations) and analyzing the spacecraft data can be prohibitive.

The requirement for autonomous operations over long periods is further complicated by two additional features of the domain: *tight resource constraints* and *hard deadlines*. A spacecraft uses various resources, including obvious ones like fuel and electrical power, and less obvious ones like the number of times a battery can be reliably discharged and recharged. Some of these resources are renewable, but most of them are not. Hence, autonomous operation requires significant emphasis on the careful utilization of non-renewable resources and on planning for the replacement of renewable resources before they run dangerously low. Spacecraft operations are also characterized by the presence of hard deadlines, e.g., the efficiency of orbit change maneuvers is a strong function of the location of the spacecraft in its orbit, so that the time at which SOI must be achieved is constrained to lie within a two hour window. Sophisticated planning and scheduling systems are needed to meet this requirement.

The second central requirement of spacecraft operation is *high reliability*. Since a spacecraft is expensive and often unique, it is essential that it achieve its mission with a high level of reliability. Part of this high reliability is achieved through the use of reliable hardware. However, the harsh environment of space and the inability to test in all flight conditions can still cause unexpected hardware failures, so the software architecture is required to compensate for such contingencies. This requirement dictates the use of an execution system with elaborate system-level fault protection capabilities. Such an executive can rapidly react to contingencies by retrying failed actions, reconfiguring spacecraft subsystems, or putting the spacecraft into a safe state to prevent further, potentially irretrievable, damage.

The requirement of high reliability is further complicated by the fact that there is *limited observability* into the spacecraft's state due to the availability of only a limited number of sensors. The addition of sensors implies added mass,[3] power, cabling, and up-front engineering time and effort. Each sensor must add clear value to the mission to be justified for inclusion. Furthermore, sensors are typically no more reliable than the associated spacecraft hardware, making it that much more difficult to deduce the true state of the spacecraft hardware. These constraints dictate the use of model-based diagnosis methods for identifying the true state of the spacecraft hardware. These methods predict unobservable state variables using a spacecraft model, and can effectively handle sensor failures.

The third central requirement of spacecraft operation is that of *concurrent activity*. The spacecraft has a number of different subsystems, all of which operate concurrently. Hence, reasoning about the spacecraft needs to reflect its concurrent nature. In particular, the planner/scheduler needs to be able to schedule concurrent activities in different parts of the spacecraft, including constraints between concurrent activities. The executive needs to have concurrent threads active to handle concurrent commands to different parts of the spacecraft. The model-based diagnosis system needs to handle concurrent changes in the spacecraft state, either due to scheduled events or due to failures.

3.2. Comparison to the Mobot Domain

The spacecraft domain shares many important features with the mobot domain, though the similarities sometimes manifest themselves in unexpected ways. There are also some important differences, both fundamentally as well as from a more pragmatic point of view.

Both mobots and spacecraft are artifacts interacting with an unengineered environment. Both have to be able to deal with unexpected contingencies, deadlines, uncertainty, and limited resources when operating autonomously.

There is also an analogy between the fundamental operations performed by mobots and spacecraft. A mobot's fundamental operation is to move from one position to another. The analogous operation on a spacecraft is to change its orientation. The time scales, degrees of freedom, and degrees of control are similar, though the geometry is different. Even obstacle avoidance has an analogous feature in spacecraft attitude control: pointing constraints to prevent sensitive instruments from pointing towards the sun. Algorithms for computing attitude trajectories are essentially path-planning algorithms in spherical coordinates. A spacecraft also moves from place to place, but the nature of ballistic trajectories and the resource constraints of realistic spacecraft make attitude control a better analogy to mobot position control. A spacecraft exercises little direct control over its position in space. It is limited to making very tiny adjustments to its velocity vector. The evolution of the velocity vector and, thus, the spacecraft's spatial destination, are mostly determined when the spacecraft is designed and launched.

These similarities are reflected in the structure of the NMRA architecture, which shares many features of the canonical "three-layer" mobot control architecture (Bonasso et al., 1997; Gat, 1992). There is a top-level architectural separation between deliberative computations, i.e., planning and scheduling (PS), a reactive decision-making executive (EXEC), and closed-loop real-time control (RT).

There are three significant differences between the spacecraft domain and the mobot domain, each of which is manifested in a particular architectural feature.

The first difference is that the source of runtime contingencies in spacecraft is usually the failure of hardware, whereas in mobots it is usually unexpected interactions with the environment. This has two important consequences for the architecture.

First, in order to properly respond to hardware failures it is important to know what hardware has failed, and this is not always immediately obvious from raw sensor data (since the sensors themselves may have failed). Thus we introduce a separate top-level component devoted to deducing the actual state of the hardware from observables. This is analogous to having a component dedicated to world modeling in a mobot, except that a significant part of the "world" being modeled is the spacecraft itself. In this sense, a spacecraft can be viewed as a mix between a mobile robot and an "immobile robot" or "immobot" as described by Williams and Nayak (1996).

The second architectural result of this difference is that the executive is structured around a single *nominal* chain of events and any deviation from nominal is considered a *failure* from which it is necessary to *recover*. This is in contrast to many mobot architectures (e.g., (Schoppers, 1987; Nilsson, 1994)), in which all possible outcomes of an action are treated equivalently by the executive, which usually just assesses the current situation without prejudice regarding the outcome of the previous action to decide what to do. This allows some mobot architectures to take advantage of serendipitous contingencies. It is a significant structural feature of current spacecraft design that unexpected contingencies are never serendipitous, though this may change in the future. If it does, it may require changes to our architecture.

The second difference between mobots and spacecraft is the degree of constraint and coupling imposed by limited resources. Terrestrial mobots usually have enough electrical and computing power that these do not have to be explicitly managed. By contrast, on a spacecraft everything is coupled to everything else through multiple mechanisms (e.g., power, thermal, and vibrational). Moreover, the costs of spacecraft dictate that all resources are utilitized to the greatest extent possible, even at the cost of added complexity due to increased interactions. Thus even mundane decisions, like switching on a camera, have to made in the context of the spacecraft's global situation. This feature manifests itself in our architecture in two ways. First, NMRA makes use of a concurrent, temporal planner and scheduler that can resolve potentially harmful interactions by allocating resources to concurrent activities over specified time periods. Some mobot architectures also have planners that coordinate system activity and resources, but many do not. Some, like 3T (Bonasso et al., 1997) and ATLANTIS (Gat, 1992) use the planner strictly to advise the executive, while others, like Subsumption (Brooks, 1986), dispense with the planner altogether. In a domain where an incorrect action can lead to mission failure far into the future, the planner assumes a much greater importance. Second, many mobot architectures resolve activity failures predominantly by making local responses (like trying another method when the first one fails). However, as noted above, even switching on a device may have negative interactions with other concurrent activities. Thus, a failure recovery sequence may need to be generated based on global considerations. To address this issue, NMRA's executive draws on the expertise of a model-based recovery expert, in addition to the procedural knowledge encoded into traditional mobot executives. This hybrid procedural/deductive executive thus extends the strengths of mobot executives to new domains of competence.

The third difference between mobots and spacecraft is the degree of reliability and robustness that is required. The opportunities for manual intervention on a spacecraft are severely restricted when compared to a terrestrial mobile robot. Typically, the only mechanism for interacting with a spacecraft is its radio communications link. These links have fairly low bandwidths and, for deep-space missions, substantial round-trip latencies. Furthermore, a mobot can almost always safely stop where it is if it needs to perform a lengthy computation to decide on its next action. A spacecraft can almost never buy time in this way. Even during cruise when there are few externally imposed deadlines, the spacecraft's attitude control loops must still operate and be properly controlled in order to prevent sensitive instruments from pointing at the sun and keep the antenna aligned with Earth. And, of course, deadlines imposed by the spacecraft's ballistic trajectory absolutely cannot be postponed. Spacecraft also tend to cost substantially more than mobots. For these reasons, autonomy software for a spacecraft must meet a much higher standard of reliability and robustness than has been the case for mobile robots.

4. Architecture Overview

Since our goal was to achieve complete autonomy for a complex domain in a limited amount of time, we chose from the outset to use a set of heterogeneous, state-of-the-art, general-purpose components that had been applied to solving specific subtasks in the domain. Hence, the main challenge was the integration of these components. These include a temporal planner/scheduler, a robust multi-threaded smart executive, and a model-based diagnosis and reconfiguration system (see Fig. 1).

In this section, we first describe how the Remote Agent (RA) is embedded in the overall flight software. Then, we provide an overview of the components of the NMRA, describe the high level operational cycle, and proceed to focus in on the details of each RA component. We conclude the section with a discussion of heterogeneous knowledge representations in the RA.

4.1. Embedded Remote Agent

The relationship between the NMRA and the flight software in which it is embedded is portrayed in Fig. 1. When viewed as a black-box, RA sends out commands to the *real-time control* software (RT). RT provides the primitive skills of the autonomous system, which take the form of discrete and continuous real-time estimation and control tasks. An example of an estimation task is the attitude determination loop, which notes the

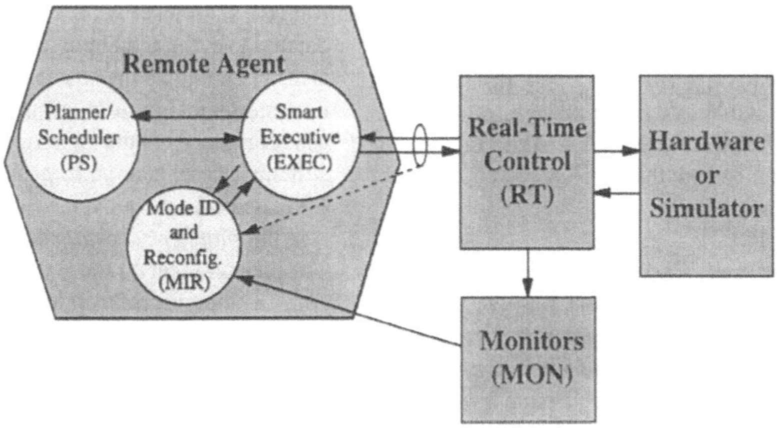

Figure 1. NMRA architecture embedded within flight software.

readings from an attitude sensor assembly (a gyroscope or a star camera) and combines it with earlier estimates to update the current estimated spacecraft attitude. An example control task is attitude control, which uses attitude effectors (a set of thrusters or reaction wheels) to change the spacecraft attitude in a way to reduce the error between commanded and estimated attitude. RT responds to high-level commands by changing the mode of a control loop or state of a device and sending a message back to RA when the command has completed.

In addition, the status of all RT control loops are passed back to RA through a set of *monitors* (MON). The monitors discretize the continuous data into a set of qualitative intervals based on trends and thresholds, and pass the results back to RA. The abstraction process is fast and simple, involving discretizing a continuous variable using thresholds on an absolute or relative scale. For example, the main engine temperature monitor has a fixed threshold above which it declares that the temperature is too high, while the inertial reference unit (IRU) monitor has a relative threshold that measures the deviation of the observed angular acceleration from the expected angular acceleration.

4.2. *RA Component Summaries*

The Remote Agent itself comprises three components: a Planner/Scheduler (PS), a Smart Executive (EXEC), and a Mode Identification and Reconfiguration component (MIR).

Planner/Scheduler (PS). PS is an integrated planner and scheduler. In our architecture, PS is activated as a "batch process" that terminates after a new schedule has been generated. It takes as input a *plan-request* that describes the current state of execution, including activities still scheduled for the future. PS combines the plan request with the goals for the current phase of the mission and produces as output a flexible, concurrent temporal plan. An output plan constrains the activity of each spacecraft subsystem over the duration of the plan, but leaves flexibility for details to be resolved during execution. The plan also contains activities and information required to monitor the progress of the plan as it is executed.

Smart Executive (EXEC). EXEC is a reactive plan execution system with responsibilities for coordinating execution-time activity.

EXEC requests a plan when necessary, by formulating a plan-request describing the current plan execution context, and then executes and monitors the generated plan. EXEC executes a plan by decomposing high-level activities in the plan into primitive activities, which it then executes by sending out commands, usually to the real-time control system. EXEC determines whether its commanded activities succeeded based either on direct feedback from the recipient of the command or on inferences drawn by the Mode Identification (MI) component of MIR. When some method to achieve a task fails, EXEC attempts to accomplish the task using an alternate method in that task's definition or by invoking the Mode Reconfiguration (MR) component of MIR as a *recovery expert.* If MR finds steps to restore the failing activity without interfering with other concurrent executing activities, EXEC performs those steps and then continues on with the original definition of the activity. If the EXEC is unable to execute or repair the current plan, it aborts the plan, cleans up all executing activities, and puts the controlled system into a stable safe state (called a *standby mode*). EXEC then requests a new plan while maintaining this standby mode until the plan is received, and finally executes the new plan.

Mode Identification and Reconfiguration (MIR). Like EXEC, MIR runs as a concurrent reactive process. MIR itself contains two components, one for *Mode Identification* (MI) and one for *Mode Reconfiguration* (MR). MI is responsible for providing a level of abstraction to the executive that enables EXEC to reason about spacecraft state in terms of a set of component modes rather than a set of low-level sensor readings. In this way, our architecture separates inferential knowledge from control knowledge. MI receives information about spacecraft state from two sources. MI obtains knowledge about the *commanded* state of the system through observing every command sent by EXEC to RT. MI obtains information of the *actual* state of the system by observing the command responses sent from RT to EXEC and from the monitoring data. MI checks the commanded state against command response and monitor data, using its declarative device models, to identify the actual mode (nominal or failed) of each spacecraft component. MI sends the inferences about the most likely mode of each component to EXEC whenever the inferred mode changes. MI also sends state updates whenever EXEC has issued a command, even if nothing changed. This enables EXEC to recognize when actions fail to have any effect at all.

MR serves as a *recovery expert* to EXEC. MR takes as input a *recovery request* from EXEC. The recovery request specifies a failed activity (or a set of activities) for which EXEC desires recovery. MR maps each activity into a set of component modes that support the activity. It compares the desired component modes to the current component modes (as inferred by MI) and then (when possible) produces a *recovery plan*. The recovery plan is a sequence of operations that, when executed starting in the current state, will move the executive into a state satisfying the properties required for successful execution of the failed activity.

4.3. *RA Operational Cycle*

Continuous autonomous operation is achieved by the repetition of the following cycle.

1. Retrieve high level goals from the mission's goals database. In the actual mission, goals can be known at the beginning of the mission, put into the database by communication from ground mission control or can originate from the operations of spacecraft subsystems (e.g., "take more pictures of star fields to estimate position and velocity of the spacecraft").
2. Ask the *planner/scheduler* to generate a schedule. The planner receives the goals, the scheduling horizon, i.e., the time interval that the schedule needs to cover, and an initial state, i.e., the state of all relevant spacecraft subsystems at the beginning of the scheduling horizon. The resulting schedule is represented as a set of *tokens* placed on various *state variable time lines*, with temporal constraints between tokens.
3. Send the new schedule generated by the planner to the *executive*. The executive will continue executing its current schedule and start executing the new schedule when the clock reaches the beginning of the new scheduling horizon. The executive translates the abstract tokens contained in the schedule into a sequence of lower level spacecraft commands that correctly implement the tokens and the constraints between tokens. It then executes these commands, making sure that the commands succeed and either retries failed commands or generates an alternate low level command sequence that achieves the token.

   In more detail, execution of a single planned activity is achieved through the following cycle:

   (a) EXEC decomposes a plan-level activity into a series of primitive activities based on execution context.
   (b) EXEC executes a primitive activity by sending a command to RT.
   (c) RT processes the command by making a change in a control loop or device state.
   (d) The monitor for the affected RT component registers the change in low-level sensor data and sends MI a new abstracted value for the state of the affected components.
   (e) MI compares the command to the observations, infers the most likely actual mode of each component, and sends an update to EXEC describing the changes in any modes of interest to EXEC.
   (f) EXEC compares the feedback from MI to the conditions specified in its task models to determine whether the command executed

successfully. If so, it proceeds to take further steps to complete the high-level activity.

(g) If EXEC receives an update from MI indicating that an activity has failed, it tries alternative methods to achieve the activity. One such method is to invoke MR as a recovery expert. In this case, the cycle is as follows:

  i. EXEC sends MR a recovery request to recover for any failed activities.
  ii. MR generates a recovery plan (when possible) consistent with the current state inferred by MI and sends the plan to EXEC.
  iii. EXEC treats the recovery plan as a new method to achieve the current activity and hence proceeds to decompose it in the same manner as other activities.

4. Hard command execution failures may require the modification of the schedule in which case the executive will coordinate the actions needed to keep the spacecraft in a "safe state" and request the generation of a new schedule from the planner.
5. Repeat the cycle from step 1 when one of the following conditions apply:

   (a) Execution (real) time has reached the end of the scheduling horizon minus the estimated time needed for the planner to generate a schedule for the following scheduling horizon;
   (b) The executive has requested a new schedule as a result of a hard failure.

We now discuss the individual components of the RA in more detail.

4.4. Planner

The goal of the NMRA planner/scheduler (Muscettola et al., 1997) is to generate a set of synchronized high-level commands that, once executed, will achieve mission goals. The NMRA planner presents several features that distinguish it from other Artificial Intelligence and Operations Research approaches to the problem. In the spacecraft domain planning and scheduling aspects of the problem need to be tightly integrated. The planner needs to recursively select and schedule appropriate activities to achieve mission goals and any other subgoals generated by these activities. It also needs to synchronize activities and allocate global resources over time (e.g., power and data storage capacity). In this domain (and in general) subgoals may also be generated due to limited availability of resources over time. For example, in a mission it would be preferable to keep scientific instruments on as long as possible (to maximize the amount of science gathered). However limited power availability may force a temporary instrument shut-down when other more mission-critical subsystems need to be functioning. In this case the allocation of power to critical subsystems (the main result of a scheduling step) generates the subgoal "instrument must be off" (which requires the application of a planning step). Considering simultaneously the consequences of planning actions and scheduling resources enables the NMRA planner to better tune the order in which decisions are made to the characteristics of the domain and, therefore, can help in keeping search complexity under control. This is a significant difference with respect to classical approaches both in Artificial Intelligence and Operations Research where action planning and resource scheduling are typically addressed in two subsequent problem solving stages, often by distinct software systems. Another important distinction between the NMRA planner and other classical approaches to planning is that besides activities, the planner also "schedules" the occurrence of states and conditions. Such states and conditions may need to be monitored to ensure that high level spacecraft conditions are correct for goals (such as spacecraft pointing states, spacecraft acceleration and stability requirements, etc.). These states can also consume resources and have finite durations and, therefore, have very similar characteristics to other activities in the plan. The NMRA planner explicitly acknowledges this similarity by using a unifying conceptual primitive, the *token*, to represent both actions and states that occur over time intervals of finite extension.

The planner used in the NMRA architecture consists of a heuristic search engine that operates in the space of incomplete or partial plans (Weld, 1994). Since the plans explicitly represent metric time, the planner makes use of a temporal database. As with most causal planners, PS begins with an incomplete plan and attempts to expand it into a complete plan by posting additional constraints in the database. These constraints originate from external goals and from constraint templates stored in a model of the spacecraft. The temporal database and the facilities for defining and accessing model information during search are provided by the HSTS system (Muscettola, 1994).

The domain model contains an explicit declaration of the spacecraft subsystems on which a token will occur. In the temporal database each subsystem has an associated timeline on which the planner inserts tokens corresponding to activities and states and resolves resource allocation conflicts. The model also contains the declaration of duration constraints and of templates of temporal constraints between tokens, called *compatibilities*. Such constraints have to be satisfied by any schedule stored in the temporal database for it to be consistent with the physics of the domain. Temporal constraint templates serve the role of generalized planning operators and are defined for any token in the domain, whether it corresponds to an activity or a state. This is a significant difference with respect to classical approaches to planning where constraint templates (also referred to as operators) are typically associated to actions but not states. The temporal database also provides constraint propagation services to verify the global consistency of the constraints posted so far.

The constraint template in Fig. 2 describes the conditions needed for an engine burn to initiate correctly (activity *Engine_Ignition* scheduled on the *(Engine Op_State)* timeline). Constraint 5 represents a request for power that increases the level of *Power_Used* on the timeline *(Power_Mgmt Power)* of an amount returned by the Lisp function call *(compute-power 'Engine_Ignition)*. Explicit invocation of external function calls provides the means for the planner to invoke "expert" modules to provide narrow but deep levels of expertise in the computation of various parameters such as durations or temperature and power levels. Access to such external knowledge is a key requirement for real-world applications of planning systems (Muscettola et al., 1995).

The planner operates by iteratively repairing "flaws" in the plan until all flaws are eliminated. There are two possible kinds of flaws:

- *Unending timeline:* a plan's timeline does not end with a token (for example, the last known token on the *(Engine Op_State)* timeline is *Engine_Idle* but the plan still allows for additional tokens to be added after it).
- *Open compatibility:* some temporal relation requested by a compatibility is still open, i.e., the planner has not selected an explicit token to satisfy it.

Figure 3 summarizes the basic "flaw fixing" loop used by the planner. To address the two possible flaw types, the planner uses the following resolution strategies. For an *unending timeline*, it takes the last token asserted on the timeline and tries to stretch it until it completely covers the rest of the timeline. If this is not possible, the temporal database will detect a temporally inconsistent plan and force backtracking. For the *open compatibility* flaw, for each temporal relation in a compatibility the planner must identify or generate a token that satisfies it. For example, Fig. 4 shows a plan with the compatibility of *Thrust(b, 200)* completely satisfied, i.e., with all temporal relations associated to two tokens. If a temporal relation in a compatibility is open, the planner can use one of three resolution strategies. It can add a token to the plan in such a way

```
(Define_Compatibility
  ((Engine Op_State) (Engine_Ignition))
   (AND
    ;; 1.  Ignition requires
    ;;      good engine pressure
    (contained_by
       ((Engine_Tanks Pressure) Good)))
    ;; 2.  Engine must have finished
    ;;    burn preparation
    (met_by
       ((Engine Op_State) (Burn_Prep)))
    ;; 3.  Engine goes into sustained
    ;;    burn state next
    (meets
      ((Engine Op_State) (Engine_Burn)))
    ;; 4.  Injector temperature must be
    ;;      good throughout
    (contained_by
       ((Engine_Injector Temp) Good))
    ;; 5.  Formula to determine
    ;;      Power consumption
    (equal
       ((Power_Mgmt Power)
          ( + (Lisp (compute-power
                         'Engine_Ignition))
               Power_Used))))
(Define_Duration_Spec
  ((Engine Op_State) (Engine_Ignition))
    ;; minimum duration
    (Lisp (compute-duration
            'Engine_Ignition :minimum))
    ;; maximum duration
    (Lisp (compute-duration
            'Engine_Ignition :maximum)))
```

Figure 2. Constraints on the *Engine_Burn_Ignition* activity.

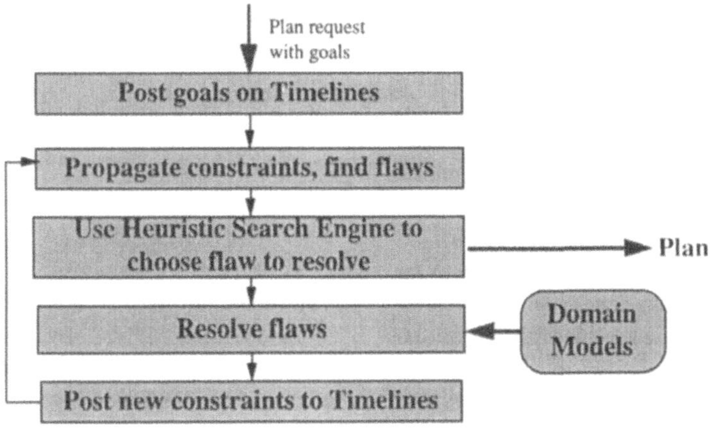

Figure 3. Basic planner cycle.

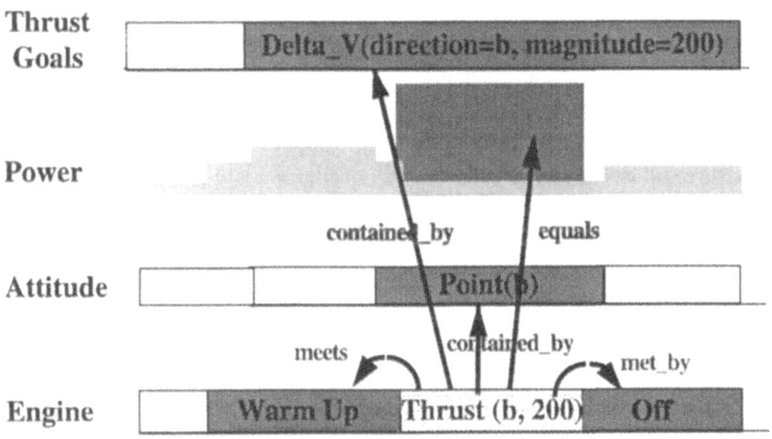

Figure 4. A plan fragment with a completely satisfied compatibility.

that it satisfies the temporal relation; it can select an existing token and impose the temporal relation on it; or it may notice that the needed token can fall outside of the time horizon covered by the plan and therefore decide to defer the satisfaction of the relation.

At each point in the search the planner must choose between several alternatives. Each choice is typically made using heuristics (e.g., "give highest priority among the flaws to be repaired to those associated to an *Engine_Burn* token") and, when heuristic information is not particularly strong, using a uniform randomized selection rule. If the wrong decision is made, PS will eventually reach a dead end, backtrack, and try a different path. Once the plan is free of flaws, the planner uses an iterative sampling approach (Langley, 1992) to heuristically improve on certain aspects of schedule quality, although it does not guarantee even local optimality along this metric. The generation of even a single plan is costly (on the order of several CPU minutes on a SPARC20 workstation) and therefore the planner needs to be called infrequently and generate plans for relatively long temporal horizons (from several hours to a week).

4.5. A Hybrid Execution Strategy

Runtime management of all system activities is performed by a hybrid of procedural/model-based execution capabilities (Pell et al., 1997). The hybrid executive's functions include execution of the top-level operational cycle, plan execution, hardware reconfiguration and runtime resource management, plan monitoring, diagnosis and fault recovery. The hybrid executive invokes the planner to help it perform these functions. The executive also controls the low-level control software by setting its modes, supplying parameters and by responding to monitored events.

In terms of runtime management of system resources, the hybrid executive performs similar functions to a traditional operating system. The main difference is that when unexpected contingencies occur, a traditional operating system can only issue a report and abort the offending process, relying on user intervention to recover from the problem. Our executive must be able to take corrective action automatically, for example in order to meet a tight orbital insertion window.

In the event of plan failure, the executive knows how to enter a stable state (called a standby mode) prior to invoking the planner, and it knows how to express that standby mode in the abstract language understood by the planner. It is important to note that establishing standby modes following plan failure is a costly activity, as it causes us to interrupt the ongoing planned activities and lose important opportunities. For example, a plan failure causing us to enter standby mode during a comet encounter would cause loss of all the encounter science, as there is not time to re-plan before the comet is out of sight. Such concerns motivate a strong desire for plan robustness, in which the plans contain enough flexibility, and the executive has the capability, to continue execution of the plan under a wide variety of execution outcomes (Pell et al., 1997).

Our executive can be viewed as a hybrid system that shares execution responsibilities between a classical reactive execution system, built on top of RAPS (Firby, 1978), and a novel mode identification and reconfiguration system, called Livingstone (Williams and Nayak, 1996). Metaphorically, the former embod ies an astronauts ability to quickly and flexibly assemble together procedural scripts into coherent control sequences, while the later embodies an engineer's ability to reason extensively about hardware and software at a common-sense level from first principles.

Within NMRA a striking result of this hybrid was that the substantial overlap in the ability of Livingstone and RAPS to perform recovery and hardware configuration tasks contributed enormously to the executive's overall robustness. For example, within RAPS it was quite natural to write a set of housekeeping procedures that encode standard rules of thumb for recovery, such as "if its broken reset it" or "if its not needed, turn it off". Meanwhile, reasoning through the model allowed Livingstone to exploit the considerable redundancy in Cassini's hardware, such as identifying novel ways of exploiting partially stuck thrusters to achieve attitude adjustments. The ability of the combined system to quickly dispense with software glitches, a demonstrator's nightmare, provided a crucial turning point in the technology's acceptance.

4.6. *Procedural Executive*

The procedural part of the hybrid executive, EXEC, is based on RAPS (Firby, 1978). RAPS provides a specialized representation language for describing context-dependent contingent response procedures, with an event-driven execution semantics. The language ensures reactivity, is natural for decomposing tasks and corresponding methods, and makes it easy to express monitoring and contingent action schemas. Its runtime system then manages the reactive exploration of a space of alternative actions by searching through a space of task decompositions.

The basic runtime loop of the executive is illustrated in Fig. 5. The system maintains an *agenda* on which all tasks are stored. Tasks are either active or sleeping. On each pass through the loop, the executive checks the external world to see if any new events have occured. Examples of events include mode updates from the mode identification system, announcements of commanded activity completion from external software, and requests from external users. The executive responds to these events by updating its internal model of the world, changing the status of affected tasks, and installing new tasks onto the agenda. It then selects some active task (based on heuristics) and performs a small amount of processing on the task. Processing a high-level task involves breaking it up into subtasks, possibly choosing among multiple methods,[4] whereas processing a primitive task involves sending messages to external software systems. At this point, the agenda is updated, and the basic reactive loop repeats.

Most architectures integrating RAPS with an explicit planner rely on the planner to drive execution by triggering invocation of RAPs one at a time, as in 3T (Bonasso et al., 1997). In NMRA, by contrast, the planner is only invoked periodically, when needed by EXEC. Hence EXEC must process and execute a large concurrent, temporal plan. We achieved this by developing a mechanism to translate a whole plan into a large task network, which is then installed as the method for a dynamically defined RAP. For each *token* in the plan, the task network contains a set of *steps*. One step maps

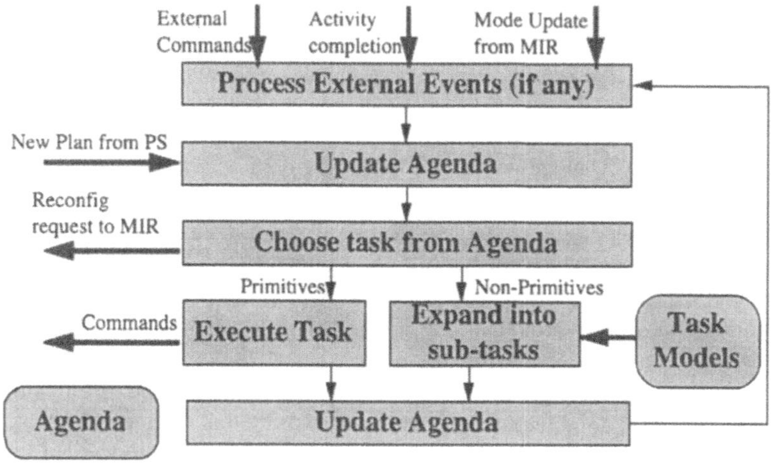

Figure 5. Executive task expansion flowchart.

directly to the name of a procedure that defines how the token itself should be executed (for example, the `take-picture` token is defined by a procedure which turns on the camera and interacts with the camera software to complete the picture). The other steps, with attached RAP annotations, are used to synchronize the token with other tokens in the plan and to ensure that execution enforces the *compatibilities* in the plan.

For example, if a compatibility in the plan says that one token, for taking a picture of a target, must be contained by another token, for pointing at that target, the translated task-network will contain steps for: (a) the pointing activity, (b) the picture activity, (c) the indication that the pointing activity has started, and (d) the indication that picture activity has finished, among others. The picture activity step will have an attached annotation stating that it cannot start until the pointing activity step has started, and the pointing activity step will have an attached annotation stating that it cannot finish until after the picture step has finished.

As it executes the plan, the entire transformed task-network is installed into EXEC's agenda. EXEC is free to work in parallel on any plan steps whose constraints are satisfied. In the example above, the annotations on the steps will force EXEC to wait until the spacecraft is pointing at the target before it takes a picture, and then to wait until the picture has finished before turning to the next target.

Since each step in the task-network is considered a subgoal of the dynamically-defined plan-execution method, the failure of a plan step (after exhausting all attempts at recovery) leads to failure of the entire plan. RAPS maintains the dependency structure and automatically removes all plan steps from the agenda, but leaves intact all other activities on the agenda. One such activity watches for plan-failure situations and installs a goal to enter standby mode and then replan, thus enforcing the top-level failure-driven replanning loop.

RAPS encourages a close adherence to a reactive programming principle of limiting deduction within the sense-act loop to that of constructing task decompositions using a limited form of matching. This ensures quick response time, which is essential to the survival of the spacecraft. Nevertheless it places a burden on the programmer of deducing, *a priori*, the consequences of failures and planning for contingencies. This is exacerbated by subtle hardware interactions, multiple and unmodeled failures, the mixture of interactions between computation, electronics and hydraulic subsystems, and limited observability due to sensor costs. These concerns are covered by the model-based component of the hybrid executive.

4.7. *Mode Identification and Reconfiguration*

The second half of the hybrid executive, the Livingstone model-based identification and reconfiguration system (MIR), complements RAPS reactive capabilities by providing a set of deductive capabilities along the sense-act loop that operate on a single, compositional model. These models permit significant on-the-fly deduction of system-wide interactions, used to process new sensor information (mode identification) or to evaluate the effects of alternate recovery actions (mode reconfiguration). Livingstone respects the intent of reactive systems, using propositional deductive capabilities (Nayak and Williams, 1997) coupled to anytime algorithms (Dean and Boddy, 1986) that have proven exceptionally efficient in the model-based diagnosis of causal systems. Through the models, Livingstone is able to reason reactively from knowledge of failure to optimal actions that reestablish the planner's primitive goals, while mitigating the failures' effects.

Livingstone also has its limitations, which are nicely met by RAPS' procedural capabilities. Livingstone's assurance of fast inference is achieved through strong restrictions on the representation used for possible recovery actions and even more severe limitations on the way in which these actions are combined (but see (Williams and Nayak, 1997) for a much more extensive model-based execution capability that preserves reactivity). One way to preserve reactivity while improving expressiveness is for a programmer or deductive system to script these complex actions before the fact. RAPS supports this, providing a natural complement to Livingstone's deductive capabilities. For example, with respect to recovery, Livingstone provides a service for selecting, composing together and deducing the effects of basic actions, in light of failure knowledge. Meanwhile RAPS provides powerful capabilities for elaborating and interleaving these basic actions into more complex sequences, which in turn may be further evaluated through Livingstone's deductive capabilities.

We now consider Livingstone in more detail. The *mode identification* (MI) component of Livingstone is responsible for identifying the current operating or failure mode of each component in the spacecraft. MI is the sensing component of Livingstone's model-based execution capability, and provides a layer of abstraction to the executive: it allows the executive to reason about the state of the spacecraft in terms of component modes, rather than in terms of low level sensor values. For example, the hybrid executive need only reason about whether a valve is open or closed, rather than having to worry about all combinations of sensor values that imply that a valve is open, and whether particular combinations of sensor values mean that the valve has failed or that a valve sensor has failed.

MI provides a variety of functions within the overall architecture. These include:

- Mode confirmation: Provide confirmation to the executive that a particular spacecraft command has completed successfully.
- Anomaly detection: Identify observed spacecraft behavior that is inconsistent with its expected behavior.
- Fault isolation and diagnosis: Identify components whose failures explain detected anomalies. In cases where models of component failure exist, identify the particular failure modes of components that explain anomalies.
- Token tracking: Monitor the state of properties of interest to EXEC, allowing it to monitor plan execution.

The *mode reconfiguration* (MR) component of Livingstone is responsible for identifying a set of control procedures that when invoked take the spacecraft from the current state, to a lowest cost state that achieves a set of goal behaviors. MR can be used to support a variety of functions within the architecture, including:

- Mode configuration: Places the spacecraft in a least cost hardware configuration that exhibits a desired behavior.
- Recovery: Moves the spacecraft from a failure state to one that restores a desired function.
- Standby and Safing. In the absence of full recovery, places the spacecraft in a safe state while awaiting additional guidance from the high-level planner or ground operations team.
- Fault avoidance: Given knowledge of current, irreparable failures, finds alternative ways of achieving desired goals.

Livingstone's MI and MR components use algorithms adapted from model-based diagnosis (de Kleer and Williams, 1987, 1989) to provide the above functions (see Fig. 6). The key idea underlying model-based diagnosis is that a combination of component modes is a possible description of the current state of the spacecraft only if the set of models associated with these modes is consistent with the observed sensor values. Following de Kleer and Williams (1989), MI uses a conflict directed best-first search to find the most probable combination of component modes consistent with the observations. This approach is independent of the actual set of available sensors, and does not require that all aspects of the spacecraft state be directly observable, providing an elegant solution to the problem of limited observability discussed in Section 3. MR uses conflict-directed best first search to identify a least cost configuration of component modes that entail a set of goal behaviors. MR only considers those combinations that are reachable from the current state,

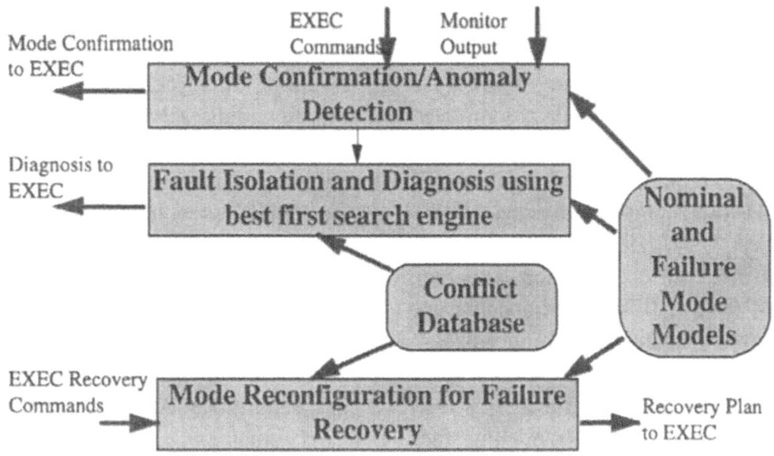

Figure 6. Architecture of Livingstone's mode identification and reconfiguration capabilities.

identified by MI, through the concurrent execution of a set of component-level control procedures. This limited ability to reconfigure component modes ensures reactivity, but precludes the generation of some types of sequences of recoveries (but see (Williams and Nayak, 1997) for an approach to overcome these limitations).

The use of model-based diagnosis algorithms as a foundation immediately provides Livingstone with a number of additional features. First, the search algorithms are sound and complete, providing a guarantee of coverage with respect to the models used (Nayak and Williams, 1997; Hamscher et al., 1992). Second, the model building methodology is modular, which simplifies model construction and maintenance, and supports reuse. Third, the algorithms extend smoothly to handling multiple faults. Fourth, while the algorithms do not require explicit fault models for each component, they can easily exploit available fault models to find likely failures.

Livingstone extends the basic modeling paradigm used in model-based diagnosis by representing each component as a finite state machine, and the whole spacecraft as a set of concurrent, synchronous state machines. Modeling components as finite state machines allows MI to effectively track state changes resulting from executive commands and allows MR to plan control sequences that move from a current to target state. Modeling the spacecraft as a concurrent machine allows MI to effectively track concurrent state changes caused either by executive commands or component failures, and allows MR to plan concurrent actions.

Another important feature of Livingstone is that it models the behavior of each component mode using abstract, or qualitative, models (Weld and de Kleer, 1990; de Kleer and Williams, 1991). These abstract models are encoded as a set of propositional clauses, allowing the use of efficient incremental unit propagation for behavior prediction (Nayak and Williams, 1997). In addition to supporting efficient behavior prediction, abstract models are much easier to acquire than detailed quantitative engineering models, and yield more robust predictions since small changes in the underlying parameters do not affect the abstract behavior of the spacecraft. Spacecraft modes are a symbolic abstraction of non-discrete sensor values and are synthesized by the monitoring module.

Finally, Livingstone uses a single model and a kernel algorithm, generalized from diagnosis, to perform all of MI and MR's functions. The combination of a small kernel with a single model, and the process of exercising these through multiple uses, contributes significantly to the robustness of the complete system.

4.8. *Heterogeneous Knowledge Representation*

One approach to developing an autonomy architecture is to seek a unified system based on a uniform computational framework. While this is an interesting goal, often the complexity of a real-world domain forces researchers to compromise on complete autonomy or to address simpler domains and applications. In our case, the challenge was to achieve complete autonomy for a very complex domain in a limited amount of time. Therefore, we chose from the outset to use a set of heterogeneous, state-of-the-art, general-purpose components that had been applied to solving specific subtasks in the domain, with the main challenge being the integration of these components. While this approach enabled us to achieve our goal of complete autonomy, it raised an important issue: the different computational engines all require different representations. These heterogeneous representations have both benefits and difficulties.

One benefit of having each computational engine look at the spacecraft from a different perspective is that the heterogeneous knowledge acquisition process aids in attaining *coverage and completeness*. Each new perspective on a subsystem potentially increases the understanding, and hence improves the modeling, for each of the other components, which also represent knowledge of that subsystem. Another benefit is *redundancy*, where overlapping models enable one component to compensate for restrictions in the representation of another component. This is particularly true in the hybrid executive, where the rich control constructs in RAPS nicely complement the deductive capabilities of Livingstone. A third benefit is *task specialization*, in which each component's representation can be optimized for solving a particular kind of task. This means that we can manually tailor each component's representations to solve problems for which it is particularly well suited.

An important example of this last point is illustrated in the representational differences between the planner/scheduler and the hybrid execution system. In NMRA the planner is concerned with activities at a high-level of abstraction, each of which encapsulates a detailed sequence of executive-level commands. A fundamental objective for the planner is to allocate resources to the high-level activities so as to provide a time and resource envelope that will ensure correctness of execution for each executive-level detailed sequence. An interval-based representation of time is suitable for this purpose. From this perspective, the planner does not really need to know if a time interval pertains to an activity or a state. However, this knowledge is crucial to ensure correct execution. The executive is interested in the occurrence of events, i.e., the transition between time intervals in the planner's perspective. To generate the appropriate commands and set up the appropriate sensor monitors, the executive needs to know if an event is controllable (the executive needs to send a command), observable (the executive expects sensory information), or neither (the executive can deduce information on the state on the basis of the domain model). Our approach localizes such distinctions to the executive's knowledge representation. This frees the planner to reason efficiently about intervals and enables us to move responsibility flexibly between other architectural components (for example, let the control tasks handle an activity which was formerly decomposed by the executive, or viceversa) without having to modify the planner's models.

While heterogeneous representations have a number of benefits, they also raise significant difficulties. Specifically, the overhead of ensuring consistency and coherence across the heterogeneous representations can be enormous. At its core, this difficulty stems from a conceptually single piece of knowledge being represented independently for each component, making it easy to introduce discrepancies. Furthermore, updating representations to reflect changes in spacecraft designs is also onerous, since the same change needs to be made (consistently) in multiple places. The traditional approach to solving this problem, and the primary approach we took, involves knowledge acquisition meetings and model review meetings involving knowledge engineering representatives for all components. However, not only are these meetings time-consuming, they are also error prone: there is rarely enough time to support in-depth reviews with all interested parties, and the resulting agreements, being in English, can lead to misunderstandings. Such errors often show up during integration and test, causing expensive schedule delays.

An alternate approach is to develop a common representation language in which each conceptually independent unit of knowledge is represented exactly once. The representations actually used by different components are automatically compiled from this single representation, thus guaranteeing consistency and coherence while retaining the benefits of heterogeneous representations discussed above. Model updates are simplified since changes need only be made in once place. We have made some progress on this front by heading toward a more unified representation of some modeled properties. First, the unified modeling for MI/MR in Livingstone (see Section 4.7) has proven to be extremely useful. Second, we use code generation techniques to translate some modeled properties, such as device power requirements, into the different representations used for each computational engine. We are working toward developing a single representation of the spacecraft model (the *one true model*, a holy grail of AI), by generalizing from the powerful heterogeneous models capable of handling the complexities of our real-world domain. We are also working on more sophisticated compilation techniques that automatically incorporate the abstractions, approximations, and reformulations needed to optimize the representation for each component.

5. Implementation

The implemented NMRA architecture successfully demonstrated planning of a nominal scenario, concurrent execution and monitoring, fault isolation, recovery and re-planning on a simulation of the simplified Cassini SOI scenario. The implementation's performance was deemed to be suitable for use on the upcoming DS-1 mission.

The planner modeled the domain with 22 parallel timelines and 52 distinct temporal constraint templates. Each template included an average of 3 temporal constraints of which an average of 1.4 constraints synchronized different timelines. The resulting schedule for the nominal scenario included 200 distinct time intervals; a schedule generated after re-planning due to engine burn interruption included 123 time intervals. The planner generated these schedules exploring less than 500 search states in an elapsed time of less than 15 minutes on a SPARC-10. Considering the computational resources available in the DS-1 mission and the background nature of the planning process, this speed is acceptable with respect to the performance needed for DS-1.

The procedural executive contained 100 RAPS with an average of 2.7 steps per RAP. The nominal schedule translated into a task net with 465 steps, making it the biggest RAP to date. The executive interacted with the underlying control loops which operated at a cycle frequency of 4 Hz. This performance level is higher than that needed to meet the requirements of the DS-1 mission.

The SOI model for the mode identification and recovery system included 80 spacecraft components with an average of 3.5 modes per component. The structure and dynamics of the domain was captured by 3424 propositions and 11101 clauses. In spite of the very large size of the model, the conflict-centered algorithms permitted fast fault isolation and determination of recovery actions. Fault isolation took between 4 and 16 search steps (1.1 to 5.5 seconds on a SPARC-5) with an average of 7 steps (2.2 seconds). Recovery took between 4 and 20 steps (1.6 to 6.1 seconds) with an average of 9.3 steps (3.1 seconds).

6. Related Work

The New Millennium Remote Agent (NMRA) architecture is closely related to the 3T (three-tier) architecture (Bonasso et al., 1997). The 3T architecture consists of a deliberative component and a real-time control component connected by a reactive conditional sequencer. NMRA and 3T both use RAPS (Firby, 1978) as our sequencer, although we are developing a new sequencer which is more closely tailored to the demands of the spacecraft environment (Gat, 1996).[5] Our deliberator is a traditional generative AI planner based on the HSTS planning framework (Muscettola, 1994), and our control component is a traditional spacecraft attitude control system (Hackney et al., 1993). We also add an architectural component explicitly dedicated to world modeling (the mode identifier), and distinguish between control and monitoring. In contrast to 3T, the prime mover in our system is the RAP sequencer, not the planner. The planner is viewed as a service invoked and controlled by the sequencer. This is necessary because computation is a limited resource (due to the hard time constraints) and so the relatively expensive operation of the planner must be carefully controlled. In this respect, our architecture follows the design of the ATLANTIS architecture (Gat, 1992).

The current state of the art in spacecraft autonomy is represented by the attitude and articulation control subsystem (AACS) on the Cassini spacecraft (Brown et al., 1995; Hackney et al., 1993) (which supplied the SOI scenario used in our prototype). The autonomy capabilities of Cassini include context-dependent command handling, resource management and fault protection. Planning is a ground (rather than on-board) function and on-board replanning is limited to a couple of predefined contingencies. An extensive set of fault monitors is used to filter measurements and warn the system of both unacceptable and off-nominal behavior. Fault diagnosis and recovery are rule-based. That is, for every possible fault or set of faults, the monitor states leading to a particular diagnosis are explicitly encoded into rules. Likewise, the fault responses for each diagnosis are explicitly encoded by hand. Robustness is achieved in difficult-to-diagnose situations by setting the system to a simple, known state from which capabilities are added incrementally until full capability is achieved or the fault is unambiguously identified. The NMRA architecture uses a model-based fault diagnosis system, adds an on-board planner, and greatly enhances the capabilities of the on-board sequencer, resulting in a dramatic leap ahead in autonomy capability.

Ahmed et al. (1994) have also worked on architecture for autonomous spacecraft. Their architecture integrates planning and execution, using TCA (Simmons, 1990) as a sequencing mechanism. However, they focused only on a subset of the problem, that of autonomous maneuver planning. Their architecture did not address problems of limited observability or generative planning.

Systems developed for applications other than spacecraft autonomy present some features comparable to NMRA. Bresina et al. (1996) describe APA, a temporal planner and executive for the autonomous, ground-based telescope domain. Their approach uses a single action representation whereas ours uses an abstract planning language, but their plan representation shares with ours flexibility and uncertainty about start and finish times of activities. However, their approach is currently restricted to single resource domains with no concurrency. Moreover, APA lacks a component comparable to MIR for reasoning about devices.

Phoenix (Cohen et al., 1989) is an agent architecture that operates on a real-time simulated fire fighting domain. The capabilities provided by the agent are comparable to those provided by the NMRA executive although many aspects of the solution seem specific to the domain and do not appear to be easily generalizable. Unlike NMRA, Phoenix's agent does not reason explicitly about parallel action execution, since actions from instantiated plans are scheduled sequentially on a single execution timeline. A notable characteristic of Phoenix is reliance on *envelopes* (Hart et al., 1990), i.e., pre-computed expected ranges of acceptability for parameters over continuous time, which are continuously monitored for robust execution.

Among the many general-purpose autonomy architectures is Guardian (Hayes-Roth, 1995), a two-layer architecture which has been used for medical monitoring of intensive care patients. Like the spacecraft domain, intensive care has hard real-time deadlines imposed by the environment and operational criticality. One notable feature of the Guardian architecture is its ability to dynamically change the amount of computational resources being devoted to its various components. The NMRA architecture also has this ability, but the approaches are quite different. Guardian manages computational resources by changing task scheduling priorities and the rates at which messages are sent to the various parts of the system. The NMRA architecture manages computational resources by giving the executive control over deliberative processes, which are managed according to the knowledge encoded in the RAPs.

SOAR (Laird et al., 1987) is an architecture based on a general-purpose search mechanism and a learning mechanism that compiles the results of past searches for fast response in the future. SOAR has been used to control flight simulators, a domain which also has hard real-time constraints and operational criticality (Tambe et al., 1995). SOAR-based agents draw on tactical plan expansions, rather than using first-principles planning as does NMRA.

CIRCA (Musliner et al., 1993) is an architecture that uses a slow AI component to provide guidance to a real-time scheduler that guarantees hard real-time response when possible. CIRCA can make tighter performance guarantees than can NMRA, although CIRCA at present contains no mechanisms for long-term planning or state inference.

Noreils and Chatila (1995) describe a mobile robot control architecture that combines planning, execution, monitoring, and contingency recovery. Their architecture lacks a sophisticated diagnosis component and the ability to reason about concurrent temporal activity and tight resources.

The Cypress (Wilkins et al., 1995) architecture combines a planning and an execution system (SIPE-II (Wilkins, 1988) and PRS (Georgeff and Lansky, 1987)) using a common representation called ACT (Wilkins and Myers, 1995). This serves as an example of a unified knowledge representation for use by heterogeneous architectural components, as we discussed in Section 4.8. Cypress is similar to NMRA, and unlike most other architectures, in that it makes use of a component for sophisticated state inference, which corresponds to NMRA's MI component. A major difference between Cypress and NMRA is our use of an interval-based rather than an operator-based planner.

Drabble (1993) describes the EXCALIBUR system, which performs closed-loop planning and execution using qualitative domain models to monitor plan execution and to generate predicted initial states for planning after execution failures. The "kitchen" domain involved concurrent temporal plans, although it was simplified and did not require robust reactions during execution.

Currie & Tate (1991) describe the O-Plan planning system, which when combined with a temporal scheduler can produce rich concurrent temporal plans. Reese & Tate (1994) developed an execution agent for this planner, and the combined system has been applied to a number of real-world problems including the military logistics domain. The plan repair mechanism (Drabble et al., 1996) is more sophisticated then ours, although the execution agent is weaker and does not perform execution-time task decomposition or robust execution.

7. Conclusions and Future Work

This paper has described NMRA, an implemented architecture for autonomous spacecraft. The architecture was driven by a careful analysis of the spacecraft domain, and integrates traditional real-time monitoring and control with constraint-based planning and scheduling, robust multi-threaded execution, and model-based diagnosis and reconfiguration. The implemented architecture was successfully demonstrated on an extremely challenging simulated spacecraft autonomy scenario. As a result, the architecture will fly as an experiment and control the first flight of NASA's New Millennium Program (NMP). The spacecraft, NMP Deep Space One (DS-1), will launch in 1998 and will autonomously cruise to and fly-by an asteroid and a comet. NMRA will be the first AI system to autonomously control an actual spacecraft.

Our immediate work for DS-1 consists mainly in acquiring and validating models of the DS-1 spacecraft and in eliciting and addressing mission requirements. To make this possible, we are working on developing better tools for sharing models across the different heterogeneous architectural components, and for model verification and validation.

Longer term, we see three major areas of research. First, our architecture could benefit from an increased use of simulation. Currently we use a simulator for development and testing the software. This could be extended to facilitate interactive knowledge acquisition and refinement, to improve projection in the planner, or to provide a tighter integration between planning and execution (Drummond et al., 1994; Levinson, 1995). Second, our architecture leaves open issues of machine learning, which could be used to tune parameters in the control system, for optimizing search control in planning, or for modifying method selection priorities during execution. Third, we see substantial benefits in having a single representation of the spacecraft, supporting multiple uses by processes of abstraction and translation. We believe that progress toward this goal is best made by generalizing from powerful, focused models capable of representing the complexities of a real-world domain.

Appendix A: Saturn Orbit Insertion Scenario Details

This appendix provides an additional level of detail concerning the SOI scenario. In particular, we first describe one possible sequence of events that meets the goals and constraints of the SOI scenario. We then describe scenarios involving different failures which might happen in the course of the nominal scenario. In both cases, the sequence of events is merely representative, as actual behavior will depend on execution context.

A.1. Nominal Scenario

- The scenario begins one day before initial Saturn periapsis.[6]
- A plan is generated onboard based on current onboard information about the state of the spacecraft, the spacecraft trajectory with respect to Saturn, the goals for the Saturn orbit insertion mission phase, and the system constraints.
- Ground controllers desire to know about the success of certain risky activities (such as the firing of pyrotechnic devices) early enough to take action if failures occur. This forces certain activities to be scheduled early, followed by communication of the results to the ground controllers on Earth.
- Science images are desired of the initial approach to Saturn and of the rings during closest approach. Limited data recorder space means that the plan should include the recorder down-load after the approach imaging and before the ring imaging.
- Power is a limited resource and engine ignition for the SOI burn occurs when power is the tightest. Non-essential equipment (e.g., science instruments and reaction wheels) must be powered off prior to engine ignition.
- Some devices need to be warmed up prior to use. Each must be turned on early enough to assure availability when needed. In addition, some devices (e.g., the gyroscope and the accelerometer) must be calibrated during scheduled low-activity periods before they will be available for use.
- In certain fault situations (e.g., gyro failure during main engine burn), it is imperative that a prompt

switch to a backup unit be possible. Otherwise the activity requiring the backup device may have to be delayed until the backup is available. If such activities are mission-critical, this could cause mission failure.

- For the critical SOI mission phase, both primary and backup units are warmed up and ready to go. The main engine is prepared for use by powering on its electronics, opening latch valves, and pre-aiming the gimbaled engine. These activities are scheduled early enough that failures allow time to switch to the backup engine.
- During SOI preparation and science collection, the spacecraft crosses the Saturn ring plane and must go to an attitude that shields the camera from ring particles.
- The spacecraft then turns to the burn attitude, main engine ignition occurs, and the spacecraft is inserted into Saturn orbit.
- After the burn, the spacecraft is returned to a safe state. Valves are closed and electronic units are powered down to cleanup the state of the spacecraft and clear the way for other activities.
- The orbit insertion burn is scheduled to end at periapsis so that science observations may take advantage of the closest approach viewing.
- The ring-plane images are down-linked to the Earth as soon as possible.
- After transmission of science and engineering data to the ground, the scenario is complete.

A.2. Failure Scenarios

The following are examples of failure scenarios that had to be handled successfully. One class of failures involved failures in the initial phase, prior to SOI, when the main engine is being prepared for the burn. These include main engine gimbals that fail stuck when being pre-aimed, and failures of the engine valve electronics, e.g., the bipropellant latch valve driver, when being powered on. The responses to these failures involve switching to the backup main engine and preparing it for the SOI burn.

A second class of failures involved failures during the SOI burn that can be recovered from without interrupting the burn. These include failures in the inertial reference unit (IRU) and a failure of the accelerometer in which it suddenly stops communicating. The response to the IRU failure is to switch to the backup IRU, which the planner has previously warmed up and readied. The response to the accelerometer communication failure is to merely stop using accelerometer data to decide when to terminate the burn, and instead to use a timer to make the decision open-loop.

A third class of failures involved failures during the SOI burn that require the burn to be interrupted. These include an engine gimbal becoming failed stuck, the main engine becoming too hot, and the accelerometer giving an acceleration reading that is much lower than expected. In all these cases, the main engine being used for the burn is potentially unusable: an engine gimbal failure means that the engine cannot be pointed correctly to achieve SOI; an overheated main engine can irreparably damage the rest of the spacecraft; low acceleration means that the main engine is not generating enough thrust to achieve SOI. While sensor failures can also explain the above symptoms, the conservative strategy dictates assuming that the failure lies in the main engine itself. Hence, in all these cases, the response is to shut down the burn and use the backup engine to retry the burn. This requires replanning with burn restart time scheduled for when all propulsion equipment has cooled down sufficiently. The duration of this new burn must also be adjusted based on the amount of burn actually accomplished in the first attempt.

Acknowledgments

The research described in this paper was carried out partly at the NASA Ames Research Center and partly at the Jet Propulsion Laboratory, California Institute of Technology, under contract with NASA. We would like to acknowledge the invaluable contributions of Gregg Swietek, Guy K. Man and Robert D. Rasmussen and their tireless promotion of spacecraft autonomy. In addition to the authors, the NMRA autonomy prototype was accomplished through the efforts of Scott Davies, Charles Fry, Robert Kanefsky, Illah Nourbakhsh, Rob Sherwood, and Hans Thomas. We thank an anonymous reviewer for careful and extensive comments which improved the paper. John Bresina, Greg Dorais, Keith Golden, Jim Kurien, and Rich Washington provided comments on drafts of this paper.

Notes

1. Multiple simultaneous hardware failures are much less likely than a single failure, so most spacecraft are only required to survive single-point failures.
2. Among other simplifications, we abstracted away some of the details of spacecraft hardware, simplified the hardware schematic by removing some instances of similar components, and by-passed

the problems involving management of redundant flight computers by restricting our demonstration to a single-cpu configuration.
3. In a spacecraft, mass directly translates to the cost of launch and the cost of carrying extra fuel to achieve all mission maneuvers.
4. A RAP definition contains a set of alternative *methods* for achieving the RAP. Each method has an associated *context*, describing conditions under which that method should be chosen, and a *priority*. The RAP interpreter uses heuristics to decide which method to execute, based on the context, priority, and historical information about the success of the different methods so far.
5. The ESL system (Gat, 1996) has now replaced RAPS as the core engine for the DS-1 Executive.
6. *Periapsis* refers to the point at which the spacecraft is closest to the planet.

References

Ahmed, A., Aljabri, A.S., and Eldred, D. 1994. Demonstration of on-board maneuver planning using autonomous s/w architecture. In *8th Annual AIAA/USU Conference on Small Satellites*.

Bonasso, R.P., Kortenkamp, D., Miller, D., and Slack, M. 1997. Experiences with an architecture for intelligent, reactive agents. *JETAI*, 9(1).

Bresina, J., Edgington, W., Swanson, K., and Drummond, M. 1996. Operational closed-loop obesrvation scheduling and execution. In *Proc. of the AAAI Fall Symposium on Plan Execution*, L. Pryor (Ed.), AAAI Press.

Brooks, R.A. 1986. A robust layered control system for a mobile robot. *IEEE Journal of Robotics and Automation*, 2(1):14–23.

Brown, G.M., Bernard, D.E., and Rasmussen, R.D. 1995. Attitude and articulation control for the cassini spacecraft: A fault tolerance overview. In *14th AIAA/IEEE Digital Avionics Systems Conference*, Cambridge, MA.

Cohen, P.R., Greenberg, M.L., Hart, D.M., and Howe, A.E. 1989. Trial by fire: Understanding the design requirements for agents in complex environments. *AI Magazine*, 10(3):32–48.

Currie, K. and Tate, A. 1991. O-plan: The open planning architecture. *Art. Int.*, 52(1):49–86.

de Kleer, J. and Williams, B.C. 1987. Diagnosing multiple faults. *Artificial Intelligence*, 32(1):97–130. Reprinted in *Readings in Model-Based Diagnosis*, Morgan Kaufmann: San Mateo, CA.

de Kleer, J. and Williams, B.C. 1989. Diagnosis with behavioral modes. In *Proc. of IJCAI-89*, pp. 1324–1330. Reprinted in *Readings in Model-Based Diagnosis*, Morgan Kaufmann: San Mateo, CA.

de Kleer, J. and Williams, B.C. (Eds.) 1991. *Artificial Intelligence*, Elsevier, Vol. 51.

Dean, T. and Boddy, M. 1986. An analysis of time-dependent planning. In *Proceedings Conference of the American Association for Artificial Intelligence*, pp. 49–54.

Drabble, B. 1993. Excalibur: A program for planning and reasoning with processes. *Artificial Intelligence*, 62(1):1–40.

Drabble, B., Tate, A., and Dalton, J. 1996. O-plan project evaluation experiments and results. Oplan Technical Report ARPA-RL/O-Plan/TR/23 Version 1, AIAI.

Drummond, M., Bresina, J., and Swanson, K. 1994. Just-in-case scheduling. In *Proc. of AAAI-94*, AAAI Press: Cambridge, MA, pp. 1098–1104.

Firby, R.J. 1978. Adaptive execution in complex dynamic worlds. Ph.D. thesis, Yale University.

Gat, E. 1992. Integrating planning and reacting in a heterogeneous asynchronous architecture for controlling real-world mobile robots. In *Proc. of AAAI-92*, AAAI Press: Cambridge, MA.

Gat, E. 1996. ESL: A language for supporting robust plan execution in embedded autonomous agents. In *Proc. of the AAAI Fall Symposium on Plan Execution*, L. Pryor (Ed.), AAAI Press.

Georgeff, M.P. and Lansky, A.L. 1987. Procedural knowledge. Technical Report 411, Artificial Intelligence Center, SRI International.

Hackney, J., Bernard, D.E., and Rasmussen, R.D. 1993. The cassini spacecraft: Object oriented flight control software. In *1993 Guidance and Control Conference*, Keystone, CO.

Hamscher, W., Console, L., and de Kleer, J. 1992. *Readings in Model-Based Diagnosis*, Morgan Kaufmann: San Mateo, CA.

Hart, D.M., Anderson, S.D., and Cohen, P.R. 1990. Envelopes as a vehicle for improving the efficiency of plan execution. COINS Technical Report 90-21, Department of Computer Science, University of Massachusetts at Amherst.

Hayes-Roth, B. 1995. An architecture for adaptive intelligent systems. *Artificial Intelligence*, 72.

IJCAI, 1997. *Proc. of the Fifteenth Int. Joint Conf. on Artificial Intelligence*, Morgan Kaufmann Publishers: Los Altos, CA.

Laird, J.E., Newell, A., and Rosenbloom, P.S. 1987. Soar: An architecture for general intelligence. *Artificial Intelligence*, 33(1).

Langley, P. 1992. Systematic and nonsystematic search strategies. In *Proc. of the 1st Int. Conf. on Artificial Intelligence Planning Systems*, Morgan Kaufmann, pp. 145–152.

Levinson, R. 1995. A general programming language for unified planning and control. *Artificial Intelligence*, 76.

Muscettola, N. 1994. HSTS: Integrating planning and scheduling. In *Intelligent Scheduling*, M. Fox and M. Zweben (Eds.), Morgan Kaufmann.

Muscettola, N., Pell, B., Hansson, O., and Mohan, S. 1995. Automating mission scheduling for space-based observatories. In *Robotic Telescopes: Current Capabilities, Present Developments, and Future Prospects for Automated Astronomy*, G.W. Henry and J.A. Eaton (Eds.), No. 79 in ASP Conf. Series. Astronomical Society of the Pacific, Provo, UT.

Muscettola, N., Smith, B., Chien, C., Fry, C., Rabideau, G., Rajan, K., and Yan, D. 1997. On-board planning for autonomous spacecraft. In *Proc. of the Fourth Int. Symp. on Artificial Intelligence, Robotics, and Automation for Space (i-SAIRAS)*, D. Atkinson (Ed.), Tokyo, Japan. Jet Propulsion Laboratory.

Musliner, D., Durfee, E., and Shin, K. 1993. Circa: A cooperative, intelligent, real-time control architecture. *IEEE Transactions on Systems, Man, and Cybernetics*, 23(6).

Nayak, P.P. and Williams, B.C. 1997. Fast context switching in real-time propositional reasoning. In *Proc. of AAAI-97*, AAAI Press: Cambridge, MA.

Nilsson, N.J. 1994. Teleo-reactive programs for agent control. *JAIR*, 1:139–158.

Noreils, F. and Chatila, R. 1995. Plan execution monitoring and control architecture for mobile robots. *IEEE Transactions on Robotics and Automation*.

Pell, B., Bernard, D.E., Chien, S.A., Gat, E., Muscettola, N., Nayak, P.P., Wagner, M.D., and Williams, B.C. 1996. A remote agent prototype for spacecraft autonomy. In *Proc. of the SPIE Conf. on Optical Science, Engineering, and Instrumentation*.

Pell, B., Gamble, E., Gat, E., Keesing, R., Kurien, J., Millar, B., Nayak, P.P., Plaunt, C., and Williams, B. 1997. A hybrid procedural/deductive executive for autonomous spacecraft. In

Procs. of the AAAI Fall Symposium on Model-Directed Autonomous Systems, P.P. Nayak and B.C. Williams (Eds.), AAAI Press.
Pell, B., Gat, E., Keesing, R., Muscettola, N., and Smith, B. 1997. Robust periodic planning and execution for autonomous spacecraft. In *Proc. of IJCAI-97*, Morgan Kaufmann Publishers: Los Altos, CA.
Pryor, L. (Ed.) 1996. *Procs. of the AAAI Fall Symposium on Plan Execution*, AAAI Press.
Reece, G. and Tate, A. 1994. Synthesizing protection monitors from causal structure. In *Procs. AIPS-94*, AAAI Press.
Schoppers, M.J. 1987. Universal plans for reactive robots in unpredictable environments. In *Procs. Int. Joint Conf. on Artificial Intelligence*, pp. 1039–1046.
Simmons, R. 1990. An architecture for coordinating planning, sensing, and action. In *Proc. DARPA Workshop on Innovative Approaches to Planning, Scheduling and Control*, DARPA, Morgan Kaufmann: San Mateo, CA, pp. 292–297.
Tambe, M., Johnson, W.L., Jones, R.M., Koss, F., Laird, J.E., Rosenbloom, P.S., and Schwamb, K. 1995. Intelligent agents for interactive simulation environments. *AI Magazine*, 16(1):15–39.
Weld, D.S. 1994. An introduction to least commitment planning. *AI Magazine*.
Weld, D.S. and de Kleer, J. (Eds.) 1990. *Readings in Qualitative Reasoning About Physical Systems*. Morgan Kaufmann Publishers, Inc.: San Mateo, California.
Wilkins, D.E. 1988. *Practical Planning*, Morgan Kaufman: San Mateo, CA.
Wilkins, D.E. and Myers, K.L. 1995. A common knowledge representation for plan generation and reactive execution. *Journal of Logic and Computation*.
Wilkins, D.E., Myers, K.L., Lowrance, J.D., and Wesley, L.P. 1995. Planning and reacting in uncertain and dynamic environments. *JETAI*, 7(1):197–227.
Williams, B.C. and Nayak, P.P. 1996a. Immobile robots: AI in the new millennium. *AI Magazine*, 17(3):16–35.
Williams, B.C. and Nayak, P.P. 1996b. A model-based approach to reactive self-configuring systems. In *Proc. of AAAI-96*, AAAI Press: Cambridge, MA, pp. 971–978.
Williams, B.C. and Nayak, P.P. 1997. A reactive planner for a model-based executive. In *Proc. of IJCAI-97*, Morgan Kaufman Publishers: Los Altos, CA.

Barney Pell is a Senior Computer Scientist in the Computational Sciences Division at NASA Ames Research Center. He is one of the architects of the Remote Agent for New Millennium's Deep Space One (DS-1) mission, and leads a team developing the Smart Executive component of the DS-1 Remote Agent. Dr. Pell received a B.S. degree with distinction in Symbolic Systems at Stanford University. He received a Ph.D. in computer science at Cambridge University, England, where he studied as a Marshall Scholar. His current research interests include spacecraft autonomy, integrated agent architecture, reactive execution systems, collaborative software development, and strategic reasoning. Pell was guest editor for Computational Intelligence Journal in 1996 and has given tutorials on autonomous agents, space robotics, and game-playing.

Doug Bernard received his B.S. in Mechanical Engineering and Mathematics from the University of Vermont, his M.S. in Mechanical Engineering from MIT and his Ph.D. in Aeronautics and Astronautics from Stanford University. He has participated in dynamics analysis and attitude control system design for several spacecraft at JPL and Hughes Aircraft, and was the Attitude and Articulation Control Subsystem (AACS) systems engineering lead for the Cassini mission to Saturn. Currently, Dr. Bernard is group supervisor for the flight system engineering group at JPL and Program Element Manager for the Remote Agent Experiment for New Millennium Program's Deep Space One mission.

Steve Chien is Technical Group Supervisor of the Artificial Intelligence Group, at the Jet Propulsion Laboratory, California Institute of Technology where he leads a multi-million dollar laboratory technology area in automated planning and scheduling for: spacecraft mission planning, maintenance of space transportation systems, science data analysis, and Deep Space Network Antenna operations. He received a Ph.D. in Computer Science in 1990 from the University of Illinois. Dr. Chien is a 1995 recipient of the Lew Allen Award for Excellence, JPL's highest honor for researchers early in their careers. In 1997 he was awarded the NASA Exceptional Achievement Medal. Dr. Chien's research interests lie in the areas of planning, scheduling, operations research, and machine learning and he has authored numerous publications in these areas.

Erann Gat is a senior member of the technical staff at the Jet Propulsion Laboratory, California Institute of Technology, where he has been working on autonomous control architectures since 1988. In 1991 Dr. Gat developed the ATLANTIS control architecture, one of the first integrations of deliberative and reactive components to be demonstrated on a real robot. ATLANTIS was used as the basis for a robot called Alfred which won the 1993 AAAI mobile robot contest. Dr. Gat was also the principal architect of the control software for Rocky III and Rocky IV, the direct predecessors of the Pathfinder Sojourner rover. Dr. Gat escapes the dangers of everyday life in Los Angeles by pursuing safe hobbies like skiing, scuba diving, and flying small single-engine airplanes.

Nicola Muscettola is a Senior Computer Scientist at the the Computational Sciences Division of the NASA Ames Research Center. He received his Diploma di Laurea in Electrical and Control Engineering and his Ph.D. in Computer Science from the Politectnico di Milano, Italy. He is the principal designer of the HSTS planning framework and is the lead of the on-board planner team for the Deep Space 1 Remote Agent Experiment. His research interests include planning, scheduling, temporal reasoning, constraint propagation, action representations and knowledge compilation.

Pandurang Nayak is a Senior Computer Scientist at the Computational Sciences Division of the NASA Ames Research Center. He received a B.Tech. in Computer Science and Engineering from the Indian Institute of Technology, Bombay, and a Ph.D. in Computer Science from Stanford University. His Ph.D. dissertation, entitled "Automated Modeling of Physical Systems", was an ACM Distinguished Thesis. He is currently an Associate Editor of the Journal of Artificial Intelligence Research (JAIR), and his research interests include model-based autonomous systems, abstractions and approximations in knowledge representation and reasoning, diagnosis and recovery, and qualitative and causal reasoning.

Michael D. Wagner received his BSE in Electrical Engineering from Duke University in 1989 and his MSE in Electrical Engineering/Artificial Intelligence from the University of Southern California in 1991. From 1989 until 1996, he served as an officer in the US Air Force, where he worked in the research and development of spacecraft sensor systems. His final Air Force assignment was as Telerobotics Representative to NASA Ames Research Center, where he developed sensor software for the Russian Marsokhod Martian rover and led the Ames team in the development of the New Millennium Autonomy Architecture Rapid Prototype (NewMAAP). In 1996, Michael co-founded Fourth Planet, Inc., a company specializing in the visualization of complex, real-time information.

Brian C. Williams is Technical Group Supervisor of the Intelligent Autonomous Systems Group at the NASA Ames Research Center, and co-lead of the model-based autonomous systems project. He received his bachelor's in Electrical Engineering at MIT, continuing on to receive a Masters and Ph.D. in Computer Science. While at MIT he developed one of the earliest qualitative simulation systems, TQA, a hybrid qualitative/quantitative symbolic algebra system, MINIMA, and a system IBIS for synthesizing innovative controller designs.

Williams was at Xerox PARC from 1989 to 1994, where he is best known for his work on the GDE and Sherlock model-based diagnosis systems. Williams received AAAI best paper awards in 1988 and 1997 for his work on qualitative symbolic algebra and incremental truth maintenance. He was guest editor for Artificial Intelligence in 1992, Chair of the AAAI Tutorial Forum in 1996 and 1997, and is currently on the editorial board of the Journal of Artificial Intelligence Research.

Autonomous Robots 5, 53–61 (1998)

Map Generation by Cooperative Low-Cost Robots in Structured Unknown Environments

M. LÓPEZ-SÁNCHEZ, F. ESTEVA, R. LÓPEZ DE MÀNTARAS AND C. SIERRA
Artificial Intelligence Research Institute IIIA, Spanish Council for Scientific Research, CSIC, Campus UAB, 08193 Bellaterra, Barcelona, Spain
maite@iiia.csic.es
esteva@iiia.csic.es
mantaras@iiia.csic.es
sierra@iiia.csic.es

J. AMAT
Automatic Control Department, Universitat Politècnica de Catalunya, c/Pau Gargallo, 5, 08028 Barcelona, Spain
amat@esaii.upc.es

Abstract. In this paper we present some results obtained with a troupe of low-cost robots designed to cooperatively explore unknown structured orthogonal environments. In order to improve the covering of the explored zone the robots show different behaviours (routine, normal and anxious) and cooperate by transferring each other the perceived environment when they meet; therefore, not all the information of the non-returning robots is lost provided that they had encountered robots that safely returned. The returning robots deliver to a host their perceived and communicated (by other robots) partial maps and the host incrementally generates the most plausible map of the environment. To perform the map generation, a fusion, completion and alignment process of the partial maps, based on fuzzy techniques, has been developed.

Keywords: map generation, fuzzy sets, autonomous robots, structured environments

1. Introduction

With the aim of exploring an environment that is unknown but easily passable, a troupe of low cost, small autonomous robots has been developed. These robots follow the already classical line of insect robots (Alami, 1993; Brooks, 1986, 1991).

The goal of these autonomous robots is to obtain partial information about an orthogonal environment during their exploration runs and afterwards, to supply this information to a host computer that will compute the most plausible map. Using this multi-robot strategy to generate a model of the environment, we expect to achieve a better efficiency than that which would be obtained based only on a single expensive robot.

The behaviour of these small autonomous robots could be seen as a metaphor of the behaviour of ants. We did not try to follow any biological model, but in some sense we were inspired by two aspects of ants: first, how they spread when looking for sources of food, and second, how they communicate when building a path from their nest to the food. Our robots have a partially random behaviour in order to increase the coverage of the environment, and they transfer each other their information when they meet. Sharing information in this way, allows the host to get the information not only from the robots that successfully return after an

*A preliminary version of this paper was presented at the First International Conference on Autonomous Agents in February 1997.

exploratory run, but also some information from those that could not return, provided that they had encountered robots that safely returned.

The next section of this paper describes the structure and the behaviour of the robots. In Section 3 we describe the statistical error analysis performed in order to know how the error intervals increase with the distance run and the number of turns. This analysis will be used to model the imprecise location of obstacles and walls by means of fuzzy techniques. Section 4 describes the fuzzy logic-based algorithms that we have developed in order to compute the most plausible map based on the partial maps perceived by the successfully returning robots. Finally, we describe the results obtained to date, we briefly point to related work and we discuss some general points.

2. Structure of Each Mobile Robot

Each robot has been designed with the aim of being small and cheap. They must have a high autonomy and be endowed with a low cost processor to memorise a map of the perceived environment. All these requirements have lead to a solution consisting on small robots with three wheels. Two of them are steering wheels having independent motors.

The robots environment perception system and the communication with the host or with other robots is based on IR impulse modulated sensors. Since some of the robots may not be able to return to deliver their map to the host, the following communication process is established: when two of them meet along their exploration run, they back-up from one to the other all the information they have acquired so far from the environment. In this way, when a robot reaches the host, it delivers both, the information acquired during its own run as well as the information obtained from other robots that it has encountered. This communication process allows to get all the information of those non-returning mini robots that had been transferred to returning ones.

2.1. Mechanical Characteristics

Each robot is 21 cm length and 15 cm wide (see Fig. 1). The 5 cm driving wheels allow to pass some small obstacles such as carpets or electrical wires. The robots can reach a maximum speed up of 0.6 m/s, and since the battery has about one hour of autonomy, each robot can do, in principle, a maximum exploration of about 2000 m.

Figure 1. Autonomous mini-robot.

2.2. Sensing Capability

Each robot is equipped with the following sensors:

- Impulse generators at each wheel for odometry.
- Five IR proximity sensors for frontal obstacles detection and for wall following.
- A proximity sensor for the detection of the terrain horizontal discontinuities.
- Safety micro switches for the protection against collisions.
- One omnidirectional IR Emitter/Receiver sensor to detect other robots and to transmit data—with a modulated FSK signal that transmits data at 9.600 bits/s.
- One IR Receiver with a scope of 100 degrees—going from the frontal axis of the robot until its right side—to situate other robots in their neighbourhood.

2.3. Navigation Strategy

The navigation system incorporated to each robot has a partially random behaviour. Although our environment has right angles and therefore, 90° turns would be enough to move into it, the robot can also make 45° turns to increase its random movement choices.

Turns are determined by the value of a probability, different probability values: $P_1 \gg P_2 \gg P_3$ correspond to three qualitatively differentiated behaviours:

- P_1 = robot with "anxious" behaviour.
- P_2 = robot with "normal" behaviour.
- P_3 = robot with "routine" behaviour.

When the robot finds an obstacle it turns in order to be parallel to it and to follow it during a random distance—that will depend on its behaviour. The turn can be done to the right or to the left based also on a probability value P_4. The robots having a probability $P_4 < 0.5$ will show a tendency to turn to the right more often than to the left, whilst the robots having a probability $P_4 > 0.5$ will behave inversely.

The different robots of the exploration troop do not show an identical behaviour. They behave in six qualitatively different ways corresponding to the different combinations of turning behaviours and turning tendencies.

During the exploration run, the robot stores its trajectory and calculates the error of its position. When this error is bigger than a previously fixed value, the robot goes back towards its starting point. In order to ensure its returning to be as safe as possible, the robot follows its own trajectory in inverted order and eliminating loops. That is, the robot deletes from its trajectory all subtrajectories starting and ending at the same point.

2.4. *Control System*

The control unit in each robot has been designed having in mind that the hardware had to be as simple as possible but, on the other hand, it had to allow achieving a behaviour sufficiently smart in order to navigate efficiently. Furthermore, the robot had to be based on a flexible hardware to allow for experimentation of navigation and control strategies. These requirements have resulted in a design which contains three different functional modules: the navigation module that generates the trajectory to be followed; the steering module that controls the motors in order to follow the generated trajectory; and the perception module that acquires information of the environment by means of IR sensors. However it is possible to replace this module by other modules adapted to different types of sensors.

The computer used to implement the navigation control unit is a 80C186 with 1 MB of RAM to store the environment map perceived by the robot and, in case it has met other robots, the respective communicated maps. These maps are represented by the sequence of those coordinate points where the robot has turned, with mention to wall detection or wall ending—that will be referred to as singular points—at these points.

The steering control module is implemented on a 80C552 and operates with high resolution since each encoder corresponds to a displacement of only 2 mm.

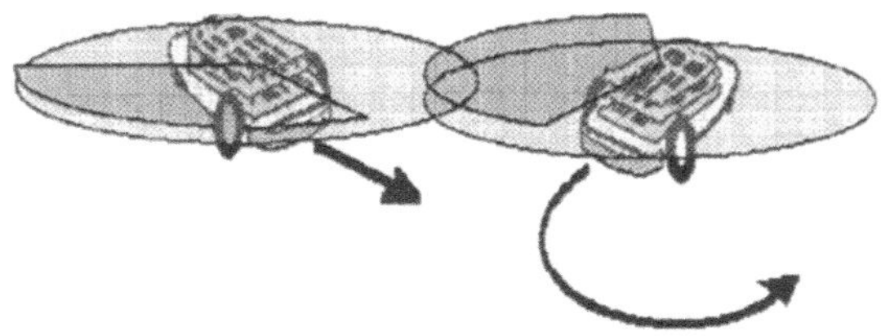

Figure 2. Two robots ending their communication process. The right one must turn left to avoid the other.

2.5. *Communication between Robots*

A robot uses its omnidirectional IR Emitter/Receiver sensor to detect another nearby robot. When this occurs, the robot changes its presence signal into an attention signal without stopping. If the other robot detects this signal, the hand-checking takes place, both robots stop and a full duplex transmission starts to communicate each other the partial map. The transmission ends in some seconds, and the robots orientate themselves to follow their previous trajectories. During this re-orientation, if a robot detects its partner with its 100° scope sensor, it avoids a possible collision by turning 90 degrees to its left (Fig. 2 depicts this situation).

If a robot meets two other robots that are already communicating each other, this robot does not detect neither the presence nor the attention signal, therefore it inhibits its presence signal in order to not interfere the communication and turns until the 100° scope sensor stops detecting any presence.

3. Error Analysis

With the goal of studying the position error of each robot due to the imprecise odometry and to the imprecise steering, we have performed an analysis based on experimental data obtained from the real robots running straight (3 and 6 m) and also turning 45 degrees left and 45 degrees right followed by a 3 m straight run. We have performed 20 trials of each run and turning situation for each robot. With the data obtained, we have used the Kolmogorov normality test to verify that the experimental sample indeed follows a normal distribution both in the direction of the trajectory and in the direction perpendicular to the trajectory and we have tested that both distributions are independent. Based on this distributions we have determined the size of an error rectangle, comprising the 95% of the sample, associated to the final position of the robot after a straight run of 3 m. This rectangle is 6.59 cm (in the direction

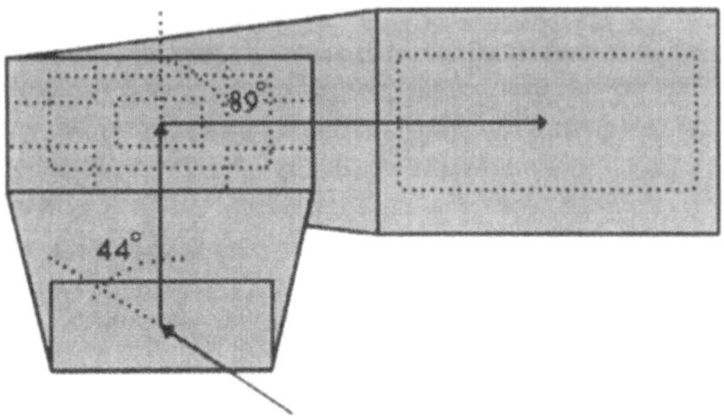

Figure 3. Error propagation.

of the trajectory) ×28.58 cm (in the direction perpendicular to the trajectory) in the average. We have also experimentally concluded that the size of the error rectangle, in a straight run, is proportional to the covered distance. Concerning the additional error due to turning, we have obtained that when robots turn 45 degrees there is, on average, an error of about ±1 degree.

3.1. Error Propagation

In free space, a trajectory is composed of a set of alternating straight runs and turns. Given the error rectangle at the initial point of a trajectory, we want to determine the error rectangle at the end of each straight runs taking into account the turning error and the error accumulated along the straight runs. Figure 3 shows the error propagation after a 45° right turn, a straight line, another 90° right turn and finally another straight line.

When following a wall, since the robot remains practically always at the same distance from the wall and since we have assumed an orthogonal environment, the error along the direction orthogonal to the wall is taken to be constant and equal to the error that the robot has after turning to place itself parallel to the wall once the wall has been detected. This error analysis and error propagation study is performed on each robot and is used by the host to model the imprecise location of the detected walls and obstacles by means of fuzzy sets as described in the next section.

4. Environment Map Generation

The goal of map generation is to obtain the most plausible position of walls and obstacles based on the assumption that the environment is orthogonal, that is, the corners between walls are angles of 90 degrees as well as the angles between any two connected sides of the obstacles. For this purpose, we represent a map as a set of wall segments. The information about the position of walls and other obstacles conveyed by the robots is imprecise due to the error in the robot position. Furthermore, different robots (or the same robot visiting twice) can detect portions of the same wall or obstacle with different degrees of imprecision. The main problem is then to decide whether several detected portions belong to the same wall or obstacle or not. This decision depends on the relative position of the segments as well as on their imprecision. The relative position of the segments can be represented by their distance. The distance is compared to a threshold to decide whether the segments represent different portions of the same wall or obstacle. If two segments are considered to belong to the same wall or obstacle a segment fusion procedure is applied to produce a single segment. This process of segment fusion is followed by a completion process in which hypothesis are made with respect to non observed regions. As we will see in Subsection 4.3, the completion process is achieved by means of a heuristic knowledge-based approach that in addition to robot information takes into account information about the structured environment in which robots evolve.

The map generation algorithm consists of three steps. The first one is the union of the current map in the host with the map perceived by a returning robot. This union is the union of both segment sets. The second step is the fusion of those segments that come from different detections but correspond to the same wall, and the algorithm ends with a completion of the resulting map.

This is an incremental approach that is necessary because, on one hand, not all troupe members may return to give the information to the host and, on the other hand, we want to have at any time the most plausible map based on the information obtained so far. Therefore, for any returning robot, the algorithm is executed to update the map. The overall algorithm is as follows:

Map-generation (RobotMap, CurrentMap):
Begin
NewMap := Fusion(CurrentMap ∪ **RobotMap)**
Return Completion(NewMap)
End

In the following sections we define the imprecise segments in terms of fuzzy sets, we give details of the segment fusion procedure and we outline the completion process.

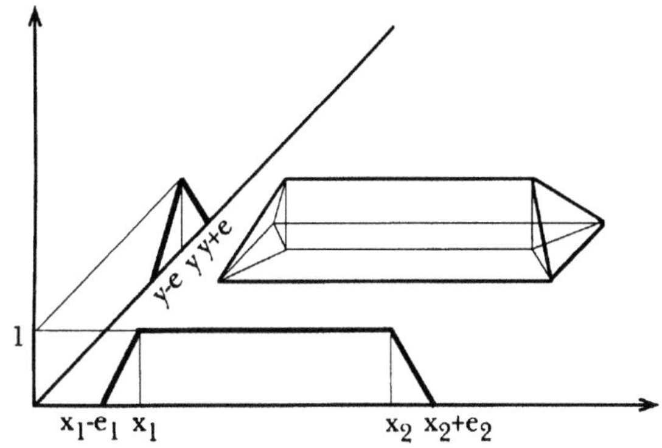

Figure 4. Horizontal imprecise segment.

4.1. *Imprecise Segments*

Imprecise segments are the basic objects of our approach, they represent wall segments—i.e., portions of walls that have been detected by the robots.

Given a detected segment $S = [(x_1, y_1), (x_2, y_2)]$, we define its associated imprecise segment $F_s : \Re \times \Re \rightarrow [0, 1]$ as the fuzzy set that gives for any point (x, y) its corresponding degree of possibility of being part of the real wall (see Zadeh, 1978; López de Màntaras, 1990 for additional details). In fact, we take a function satisfying $F_s(x, y) = 1$ iff $(x, y) \in S$ and, from the value 1 at the points of S, F_s decreases linearly untill the value 0 at the borders of the error rectangle associated to S. This rectangle has length $e_1 + S + e_2$ being e_1 and e_2 the errors in the direction of displacement, and width $2e$ along the direction orthogonal to the trajectory (see Fig. 4).

Observe that, since the distance from the robot to a wall, while following it, remains practically constant, the error, e, in the direction orthogonal to the trajectory remains constant as well. However, as we have explained in the error analysis section, the error along the direction of the trajectory increases with the distance travelled. That is, assuming in Fig. 4 a movement from left to right, we would have in fact that the error e_2 on the right would be higher than the error e_1 on the left of the figure. Furthermore, for each segment we keep the list of coordinates of the singular points that have been detected (i.e., wall endings: corners and gaps in real walls).

Given the orthogonality of the structured environment, we have only two possible orientations for any segment: vertical—where $x_1 = x_2$—or horizontal—with $y_1 = y_2$. Thus, when the segment S is $[(x_1, y), (x_2, y)]$ and the errors are e_1 and e_2 in the direction of the trajectory and e in the orthogonal direction, the function F_s is determined by the error rectangle which justify the use of the following representation of imprecise segments: $((x_1, y_1), (x_2, y_2), e, e_1, e_2)$. Actually, what we know is that there is a portion of a real wall that is represented by the segment $[(a, b), (c, b)]$ being a, b, c values satisfying $y - e < b < y + e$, $x_1 + e_1 > a > x_1 - e_1$ and $x_2 - e_2 < c < x_2 + e_2$ (and similarly for the orthogonal case). Figure 4 shows an example of the fuzzy modelling of an imprecise horizontal segment.

4.2. *Segment Fusion*

As we have said, the main problem is to decide whether several imprecise segments belong to the same wall or obstacle or not. This decision depends on the relative position of the imprecise segments like, for example, the case shown in Fig. 5(a). This relative position is determined by two parameters Δx and Δy that represent the minimum distance with respect to the two axis. For example, in the case of the two parallel vertical segments of Fig. 5(a) we have:

$$F_s : ((x, y_1), (x, y_2), e, e_1, e_2)$$

$$F'_s : ((x', y'_1), (x', y'_2), e', e'_1, e'_2)$$

$$\Delta x(F_s, F'_s) = |x - x'|$$

$$\Delta y(F_s, F'_s) = \begin{cases} 0 & \text{If } (y_1 - y'_2) * (y_2 - y'_1) < 0 \\ \min(|y_1 - y'_2|, |y_2 - y'_1|) & \text{otherwise} \end{cases}$$

In the Δy definition, the first part stands for the case in which both segments overlap. In that case the terms of the product have a different sign.

Given a map M, represented as a list of segments ordered by imprecision, the "Fusion" function computes

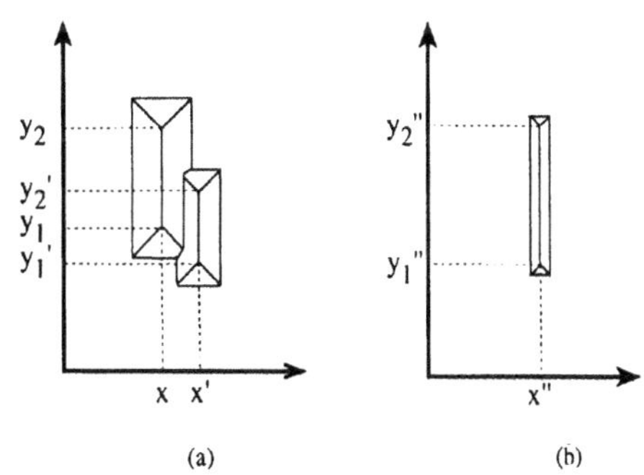

Figure 5. Vertical segments in (a) are merged into the segment in (b).

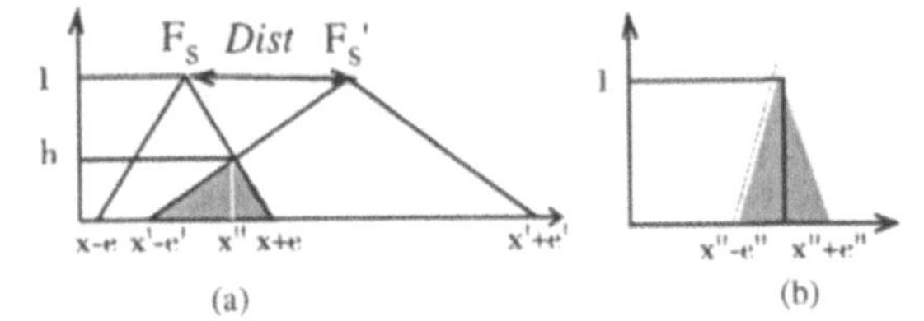

Figure 6. Merging of two vertical segment projections along the x-axis.

a new map M' whose difference with M is that some segments have been fused. For each horizontal (resp. vertical) segment $F_s \in M$, starting with the most precise, we compute its distance to the other horizontal (resp. vertical) segments in M with respect to both axis. Then, with those segments in M such that both distances are below a threshold, the segment $F_s' \in M$ with the lowest average distance with respect to both axis is selected for fusion. Next we merge F_s and F_s' and we replace them by the resulting merged segment F_s'' that is included in the list M in the appropriate place according to its error. Furthermore, the singular points are updated in order to keep the singular points of the merged segment properly connected with other segments meeting at these singular points. When no more fusions are possible, the process stops. The complexity of this algorithm is $O(n^2)$ where n is the number of segments.

Figures 5 and 6 show an example of vertical segments fusion. The coordinates y_1'' and y_2'' of the fused segment F_s'' are obvious and e_1'' and e_2'' are the corresponding errors of the previous segments. The coordinate x'' of F_s'' and its corresponding error e'' is calculated as the centre of gravity of masses $1/e$ and $1/e'$ concentrated at coordinates x and x' respectively. An easy computation shows that x'' is the first coordinate of the crossing point of the segments $[(x, 1), (x+e, 0)]$ and $[(x'-e, 0), (x', 1)]$ that are the sides of the triangle projections of the imprecise segments F_s and F_s' on the x-axis (see Fig. 6). On the other hand, e'' satisfies $1/e'' = 1/e + 1/e'$. Other alternatives for computing segment fusion are possible.

4.3. *Global Environment Map Completion*

The fusion procedure is an initial step that merges segments in a local basis using only geometrical relations between them. To improve the map, we perform a completion process based on: (1) a global view of the environment, including the map that each robot has got from the other robots that it has encountered during the exploration; (2) knowledge about the characteristics of the environment; and (3) the paths followed by the robots. We adopt an heuristic knowledge-based approach based on a set of rules covering the different situations that have been identified. These rules are divided into the following four rule sets:

- extending single segments
- combining two segments
- completing corners (L shaped and T shaped)
- completing doorways

and are considered in this order.

One example of heuristic rule for combining two vertical segments is:

If
  segments s_1 and s_2 are parallel and
  the closest extremes of s_1 and s_2
  are not singular points and
  $\Delta x(s_1, s_2) < \alpha$ and
  $\Delta y(s_1, s_2) < \beta$ and
  there is no path through the gap
  between s_1 and s_2

Then Do *Vert_Merging*(s_1, s_2)

Where α and β are decision threshold parameters: α is the typical width of a doorway and β the typical width of a corridor. *Vert_Merging* is the function that takes care of the merging.

Next we give an example of a rule for completing "L-shaped" corners:

If
  s_1 and s_2 are orthogonal and
  the closest extremes of s_1 and s_2
  are not singular points and
  $\Delta x(s_1, s_2) < \alpha$ and
  $\Delta y(s_1, s_2) < \beta$ and
  there is no path through the gap
  between s_1 and s_2 and
  s_1 and s_2 do not meet at the
  corner point (x, y)

Then
  make s_1 and s_2 to meet at (x, y) and
  label (x, y) as singular point

Similar rules are used to extend single segments, to combine two segments in other situations and to complete corners as well as doorways.

The very last step is to align segments that do not have the same x coordinate (vertical segments) or the same y coordinate (horizontal segments) but should, because, for instance, they are at both sides of a doorway. To do this alignment we simply apply the *Vert_Merging* (resp. *Hor_Merging*) functions of the fusion process in order to compute the x (or y) coordinate of the aligned segments. These Merging functions are suitable because they weight the imprecision of the two segments in such a way that the aligned position is closer to the more precise of the two segments.

5. Results

Three prototypes of autonomous robots have been physically built and tested in an orthogonal environment that has been properly arranged: it is empty of non-orthogonal obstacles and all walls are white in order to have a constant IR sensor information. Figure 7 shows this environment.

After a random walk, each robot returns to the host computer and delivers its information. The host then fuses each new partial map with the current global map. Figure 8 shows the status of the global map when the information from the two first robots has been already fused, plus the partial map delivered by the third robot. Notice that fuzzy segments have been represented as lines corresponding to the positions with possibility value equal to 1 (the error extent is not shown in this figure for the sake of clarity, since most lines that are close are in fact intersecting fuzzy segments). Those small circles at the end of some segments represent detected singular points. Finally, we can see in Fig. 9 the result of fusing the last partial map with the global map. Obviously, the coverage of the environment can be improved with further exploration of the robots.

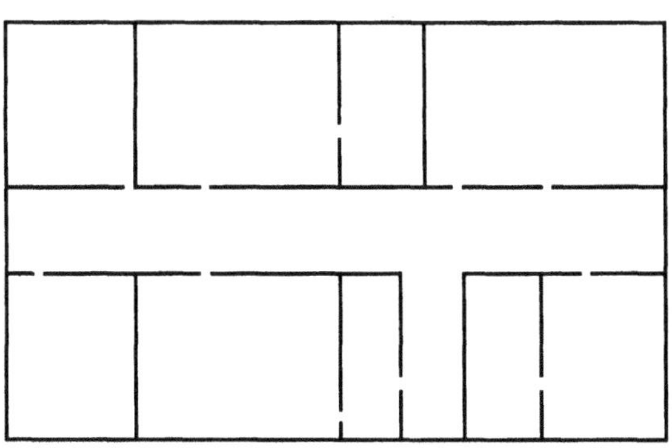

Figure 7. Example of environment.

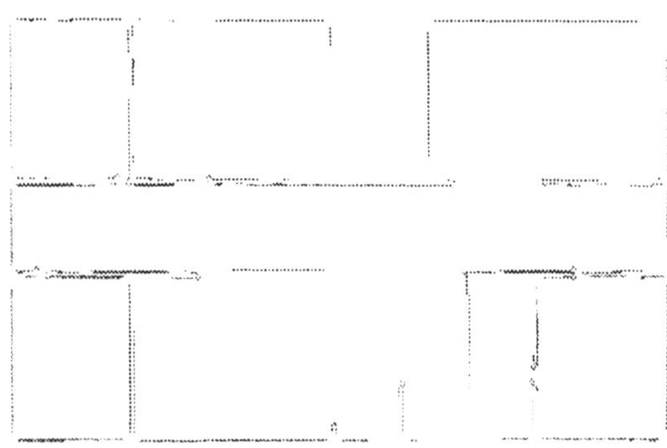

Figure 8. Map representing the union between the current map—after fusing the map from two returning robots—plus another map from a third robot.

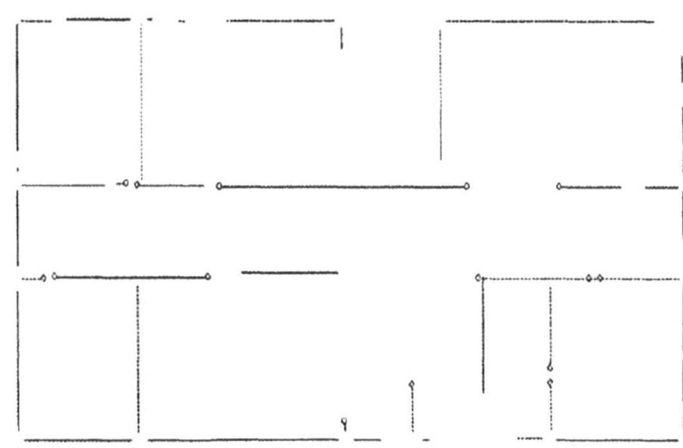

Figure 9. Final map obtained after the fusion of the map of the third robot.

6. Relation to Other Work

Two main characteristics differentiates our work from the other ones addressing the problem of map building: the kind of sensors that are used and the techniques that are applied for map representation.

Concerning the map representation, the use of probabilistic techniques is quite common, for example (Betge-Brezetz et al., 1996) define landmarks in natural environments assuming a gaussian certainty distribution of their positions and (Elfes and Moravec, 1985; Lim and Cho, 1992; Pagac et al., 1996) estimate the probability of cell occupancy in certainty grid representations. Probabilistic techniques need a big amount of data and following a known distribution, so fuzzy set theory is a good alternative when these conditions are not met: (Kim et al., 1994) use fuzzy numbers to model the uncertainty of the parameters of geometric primitives and coordinate transformations used to describe natural environments and (Poloni et al., 1995) use fuzzy sets to assign to each point in the grid map a degree of being empty and of being occupied. This last work is the closest to ours, but our use of proximity infrared sensors in short distances (i.e., 30 cm) give better results than their imprecise information coming from ultrasonic sensors. In fact, most of the works mentioned here use ultrasonic sensors for object detection. These sensors have problems that increase the location error: due to the wide radiation angle, the uncertainties on angular location are too large; the distance of detection is affected by false reflections—the beam may hit the receiver after bouncing with several obstacles—and furthermore, the ultrasonic beam can easily be lost also due to reflection when hitting obstacles with a large angle of incidence.

7. Discussion

In this paper we have presented an approach based on fuzzy segments to generate maps of unknown orthogonal environments. This task is achieved in two stages: first, a troupe of cooperative autonomous robots performs a random exploration of the environment; and second, a computer host generates a global map from the partial maps delivered by the robots.

Concerning the cooperation between robots, it is done by transmitting their partial maps when they meet. The hardware that supports this process consists on infrared emitters/receivers. They are cheap and simple enough to fulfill the restrictions of cost and size of our robots. The use of radio systems would have meant a more complex communication protocol and a higher cost.

Maps are represented by a set of fuzzy segments, that is, we propose a feature-based representation of the environment. Since we have the orthogonality assumption, we only consider horizontal and vertical segments. Both fusion and completion processes are so dependent of this characteristic that part of their success relies on it. When relaxing this condition, a discretized—grid-based—approximation is far more suitable. In fact, we have already developed this idea in (López-Sánchez et al., 1997).

Another question that we have addressed is the problem of data inconsistency, that is the case, for instance, when there is a robot trajectory going through a wall segment in the global current map. In this situation we solve the contradiction by keeping the segment in the map only if its associated error is lower than the error of the trajectory.

References

Alami, R., Chatilla, R., and Espiau, B. 1993. Designing an intelligent control arrchitecture for autonomous robots. In *Proc. 7th Int. Conf. on Advanced Robotics*, Tokyo.

Betgé-Brezetz, Hébert, Chatila, and Devy 1996. Uncertain map making in natural environments. In *Proc. IEEE: Int. Conf. on Robotics and Automation*, pp. 1048–1053.

Brooks, R.A. 1986. A robust layered control system for a mobile robot. *IEEE Journal of Robotics and Automation*, RA-2(1):14–23.

Brooks, R.A. 1991. Intelligence without reason. In *Proc. Int. Joint Conf. on Artificial Intelligence*, Sydney, pp. 569–595.

Elfes, A. and Moravec, H.P. 1985. High resolution maps from wide angle sonar. In *Proc. IEEE Int. Conf. on Robotics and Automation*, St. Louis, MO, pp. 116–121.

Hager, G.D. 1990. Task-directed sensor fusion and planning: A computational approach. *International Series in Engineering and Computer Science*, Kluwer Academic Publishers: MA, Vol. SECS 99.

Hwan, J., Dong, L., and Cho, W. 1992. Physically based sensor modelling for a sonar map in a specular environment. In *Proc. IEEE Int. Conf. on Robotics and Automation*, Nice, France, pp. 1714–1719.

Kim, Hyup Ko and Jin, Chung 1994. Uncertain robot environment modelling using fuzzy numbers. *Fuzzy Sets and Systems*, 61:53–62.

Kuc, R. and Siegel, M.W. 1987. Physically based simulation model for acoustic sensor robot navigation. *IEEE Trans. Pattern Analysis Machine Learning*, 9(6):766–778.

Lim, Cho 1992. Phisically based sensor modelling for sonar map in specular environment. In *Proc. IEEE: Int. Conf. on Robotics and Automation*, pp. 1714–1719.

López de Màntaras, R. 1990. Approximate reasoning models. Ellis Horwood Series in Artificial Intelligence, UK.

López-Sánchez, López de Màntaras, and Sierra 1997. Incremental map generation by low cost robots based on possibility/necessity grids. In *Proc. Uncertainty in Artificial Intelligence*, pp. 351–357.

Pagac, Nebot, and Durrant-White 1996. An evidential approach to probabilistic map-building. In *Proc. IEEE: Int. Conf. on Robotics and Automation*, pp. 745–750.

Poloni, M., Ulivi, G., and Vendittelli, M. 1995. Fuzzy logic and autonomous vehicles: Experiments in ultrasonic vision. *Fuzzy Sets and Systems*, 69:15–27.

Zadeh, A. 1978. Fuzzy sets as a basis for a theory of possibility. *Fuzzy Sets and Systems*, 1:3–28.

Maite López-Sánchez is a Ph.D. student in the Artificial Intelligence Research Institute (IIIA) of the Spanish Scientific Research Council (CSIC). In 1994 she received B.Sc. degree in Computer Science from the Autonomous University of Barcelona. In 1996 she completed the thesis of her M.Sc. degree on Computer Science from the same university. She has other publications related to this paper and her main interests are the application of Artificial Intelligence techniques to Robotics, focusing on behaviour-based architectures, cooperation policies and case-based learning for environment modeling.

Francesc Esteva received his B.Sc. in Mathematics in 1969 and Ph.D. in Mathematics in 1974, both from the University of Barcelona. He was Full Professor at the Technical University of Catalunya and he is currently the Director of the Artificial Intelligence Research Institute (IIIA) of the Spanish Research Council (CSIC). He has published about 100 papers in Journals and Conferences mainly in the area of approximate reasoning and its applications to different areas like knowledge-based systems and case-based reasoning. Multi-valued Logics, Modal and Multi-modal Systems and mainly Fuzzy Logic other the topics he has been working on. He is currently the President of the Spanish Association for Fuzzy Logic and Technology and he is one of the members of the editorial board of the forthcoming Handbook of Fuzzy Computation published by Oxford University Press. Similarity-based reasoning is his most recent area of interest and he is co-editor of a forth comming book about this subject.

Ramon López de Màntaras is Research Full Professor and Deputy Director of the Artificial Intelligence Institute of the Spanish Higher Council for Scientific Research. He received Ph.D. in Applied Physics (Automatic Control) in 1977 from the University of Toulouse (France), M.Sc. in Engineering (Computer Science) from the University of California at Berkeley and Ph.D. in Computer Science in 1981 from the Polytechnical University of Catalonia. He has held a Visiting Professor appointment at the University of Paris VI in 1986. From 1981 to 1985 he taught at the Computer Science Department of the Polytechnical University of Catalonia. He is a member of the editorial board of twelve international journals and former Editor-in-Chief of Artificial Intelligence Communications. He is a recipient of the "City of Barcelona" Research Prize in 1982, the "European Artificial Intelligence Research Paper Award" in 1987 and the "Swets & Zeitlinger Distinguished Paper Award" of the International Computer Music Association in 1997 among other awards. Presently working on the relationship between Similarity Logics and Case-Based Reasoning, on the use of Dynamic Logic to formalize Reflective Architectures, on the problem of dealing with Uncertainty in Autonomous Robots, on Multi-Agent Systems and on Case-Based Reasoning applications to Music. He is author or co-author of over 100 scientific papers and of the book "Approximate Reasoning Models" published in the Ellis Horwood Artificial Intelligence Series.

Carles Sierra is a researcher at the Spanish Scientific Research Council. He received a B.Sc. on Computer Science in 1986 and a Ph.D. on Computer Science in 1989, both from the Technical University of Catalonia in Barcelona. He has been active in the areas of approximate reasoning and knowledge-based systems—he received the Digital award for the best paper on AI in 1987 for a work on uncertainty. Several applications of knowledge-based systems, most on medicine, have been developed and transferred under his supervision. His current research interests include the fields of multiagent systems and robotics. He has published more than 70 articles in international journals and conferences.

Josep Amat was born in Barcelona in 1940. He received his degree in Electrical Engineering in 1967 from The School of Engineering of Barcelona, and obtained his Ph.D. in 1977 from the Technical University of Catalonia. In 1968 he joined this university to teach Digital Systems. Now, he is professor in Computer Technology at the Automatic Control Department of the UPC. His research work initially was in the field of Digital Signal Processing and since 1980 his work is oriented to the Computer Vision field, for Robotics applications, mainly in the development of high speed image processors. He leads the research in this field in the Department of Automatic Control at the UPC. He is the author of many publications and has participated in different national and international conferences and symposia. He has also participated in several projects related to perception and multisensorial fusion and Underwater Robots.

Autonomous Robots, 5, 63–77 (1998)

Grounding Mundane Inference in Perception *

IAN DOUGLAS HORSWILL
The Institute for the Learning Sciences, Northwestern University, 1890 Maple Avenue, Evanston, IL 60201
ian@cs.nwu.edu

Abstract. We describe a uniform technique for representing both sensory data and the attentional state of an agent using a subset of modal logic with indexicals. The resulting representation maps naturally into feed-forward parallel networks or can be implemented on stock hardware using bit-mask instructions. The representation has "circuit-semantics" (Nilsson, 1994, Rosenschein and Kaelbling, 1986), but can efficiently represent propositions containing modals, unary predicates, and functions. We describe an example using Kludge, a vision-based mobile robot programmed to perform simple natural language instructions involving fetching and following tasks.

Keywords: vision, active perception, reasoning, knowledge representation, agent architectures

1. Introduction

Suppose you want to program a robot to accept simple instructions like "bring the green ball here." A likely plan for solving this task might be:

1. Search for a green ball and track it visually as the ball and/or robot move
2. Drive to the ball
3. Grab the ball
4. Drive to the starting position
5. Release the ball

Executing such a plan robustly requires the robot to perform error detection and recovery. If the ball slips out of the gripper during transit, the robot needs to detect the situation and rerun steps 2 and 3. Unfortunately, the plan doesn't say that the ball needs to stay in the gripper, just that the robot should perform a grasp at a certain point in the plan.

This paper addresses two problems in autonomous agency. The first is the problem of making mundane inference and problem solving processes, such as determining how to deliver a ball, run fast enough that they are effectively instantaneous. If the problem solver runs fast enough, then error detection and recovery are easy: we just keep rerunning the problem solver from scratch. The second problem is grounding the problem solver's inferences in continually updated sensor information so that decisions about what to do next change in real time as sensor readings change.

I will argue that these two problems are related. I will present an architecture that allows a useful subset of modal logic to be compiled into fast feed-forward networks. The networks both drive and are driven by a modern active vision system on a real mobile robot (figure 1). The architecture allows designers to build systems with the performance characteristics of behavior-based systems (Matarič, 1997), while simultaneously providing much of the generativity of traditional symbolic problem solvers.

1.1. *Agent architectures*

Much of the work in this decade on agent architectures has been concerned with the problem of merging the

*A preliminary version of this paper was presented at the First International Conference on Autonomous Agents in February 1997.

strengths of reactive and deliberative systems (see for example,(Hexmoor, Horswill and Kortenkamp, 1997, Hexmoor, Horswill and Kortenkamp, 1997)).[1] A pure reactive system contains no internal state (memory) and consists of a set of task-specific, pre-programmed rules for firing actions based on immediate sensory input. The rules are typically implemented using some kind of parallel network whose inputs are driven by sensors and whose outputs drive the effectors. Purely deliberative systems (planners) compute in advance, and commit to, a complete series of actions intended to achieve the goal. The series of actions (the plan) is typically computed from the goal using a combination of search and simulation.

Both approaches have obvious flaws. It is easy to construct examples of non-Markovian environments in which a reactive system's lack of internal state will get it into trouble. The task-specificity of their rules also raises questions about scaling and generativity. On the other side, deliberative systems typically require exponential search, making them slow for simple problems and unusable for complex ones. Another issue is that purely deliberative systems commit to a plan in its entirety. The only way they can respond to unexpected contingencies is for the plan to fail entirely, triggering a restart of the planner. However, it can be difficult for the executive running the plan to know how to judge when the plan has failed, since the plan contains only the actions to perform, not their rationale. In a simple planner/executive architecture, all the domain knowledge is in the planner, but understanding whether a plan has failed requires at least as much knowledge as formulating it in the first place.

Fig. 1. Kludge the robot during a delivery task. Its job is to search for, approach, and grasp a ball of specified color and to deliver it to a person wearing a specified color or to a designated point in space. It must quickly recover from problems such as the ball slipping out of its mandibles or the ball being momentarily occluded.

Clearly, one wants a system that combines the strengths of reactive and deliberative systems. The most common approach to this has been to build a hybrid system incorporating reactive and deliberative systems as components, typically in a three-tiered architecture: a planner computes plans from goals, an executive sequences them, and a set of reactive systems implement the low level symbolic actions (see (Arkin, 1997) for an extensive survey). One common argument for such an architecture is that deliberation will always be slower than reaction, so any division of labor between the two will have to allocate fast timescale activities to the reactive system and slower timescale activities to deliberative systems (Bonasso et al., 1997).

The other major approach is to select special cases of planning that can be mapped efficiently into parallel hardware. The networks then compute the same input/output mapping as a planner, but run in bounded time. The earliest example of this is Rosenschein and Kaelbling's system, which compiles propositional logic axioms into sequential circuits (Rosenschein and Kaelbling, 1986). Kaelbling's GAPPS system (Kaelbling, 1988) used goal regression to compile a (propositional) planner-like formalism into sequential circuits. Mataric implemented a Dijkstra-like shortest path finder using spreading activation in a behavior-based system (Matarič, 1992). Maes' behavior network system computed an approximation to propositional STRIPS planning using spreading activation (Maes, 1989).

The use of propositional logic (logic without variables or predicate/argument structure) is severely limiting, however, and there have been a few attempts to rectify it. Nilsson's TRT system (nilsson:teleoreactive) handles predicate/argument structure by incrementally growing a propositional network as new combinations of arguments are encountered. Agre and Chapman's *deictic representation* (Agre and Chapman, 1987, Agre, 1988) is an explicit attempt to overcome the limitations of propositional representations in which variable binding is performed in the perceptual system rather than the reasoning system.

1.2. Representation languages, complexity, and parallelizability

The problem with hybrid systems is complexity. Technically, planning isn't slow, it's *complex*. Planning systems typically take exponential time to compute their plans. If we imagine a planner that takes time exponential in the length of the completed plan, then while it might take 100 seconds to compute a 30 step plan, 3 seconds to compute a 25 step plan, and 3 microseconds to compute a 5 step plan, it would take 28 hours to compute a 40 step plan.[2]

The complexity problems of the parallel hybrids have more to do with hardware than time. Ultimately, the problem with the parallel hybrids is that they have no way of doing variable binding and so they are effectively limited to representations that are isomorphic to propositional logic rather than predicate (e.g. first order) logic. That means that instead of having a single node in the system that represents $on(A, B)$, they need to use separate nodes for each possible value of A and B. In the case of Maes' solution for the blocks world (Maes, 1989), that ends up meaning $O(n^3)$ nodes in the system and $O(n^4)$ wires connecting them. In Hasegawa et al.'s dialog system (Hasegawa et al., 1997), a separate node is used for every possible sentence that can be make by either speaker.

There is a deep relationship between the expressiveness of a representation language and the computational complexity of its inference problem (see, for example, (Levesque and Brachman, 1985)). Propositional inference is quite tractable but so limited that one is forced to do things like represent all ground-instances of $on(A, B)$ separately. On the other hand, inference in first-order predicate logic is Turing-complete and so wildly intractable. If we remove certain features like term expressions from the language (so we can say $related(X, Y)$ but not $related(brother(X), X)$), we get the language DATALOG, whose inference problem is P-complete (Ullman and Wisdom, 1997), meaning it is tractable, but so far as we know, unparallelizable.

In fact, parallelizing inference is extremely difficult. Even simple problems like unification, which are ubiquitous in AI systems, are P-complete and so thought to be inherently serial (Dwork, Kanellakis and Mitchell, 1984).

1.3. Perception systems

Contemporary active sensing systems typically have attentional components that must be controlled appropriately for the current task. For example, many active vision systems have the ability to track a small number of objects (typically one) (Inoue, 1993, Brown, Coombs and Soong, 1992, Hager, 1995, Fairley, Reid and Murray, 1995, Crisman, 1992, Lowe, 1992, Huttenlocher, Noh and Rucklidge, 1992). Attentive tracking is necessary because it is impractical to completely analyze a stream of images — find and label all objects in each image, then match the identities of the objects between images — which would amount to tracking every possible object at once. Instead, tracking systems "latch on" to a small number of objects and continuously report their characteristics over time. For example, a disparity-based tracker such as (Brown, Coombs and Soong, 1992) reports the image-plane coordinates of the object being tracked, as well as its disparity (a measure of distance). Trackers generally provide some way of initially choosing an object based on its visual properties (Prokopowicz, Swain and Kahn, 1994, Horswill, 1995).

Control of modern perceptual systems therefore involves dynamic allocation of scarce sensory resources to task-relevant objects. To know the distance of an object, the system must first allocate a tracker to the object. Unfortunately, current reasoning systems have no way of directing these resources. They require all knowledge to be pre-stored in a symbolic database. This leaves the perceptual system in the position of having to guess what information might be useful to the reasoner.

Allocating perceptual resources intelligently requires that the agent have at least limited ability to represent its own attentional state and to reason about that state's relation to its present goals. This requires care, however, since the allocation of attention cannot depend on gathering information that would itself require allocating attention (on penalty of infinite regress).

In practice, symbolic architectures typically treat the allocation of a tracker or the checking of its output as a discrete action that is generated by the problem solver based on *ad-hoc* rules. In addition to being an added burden on the programmer, such rules leave open the possibility that the problem solver will forget to check something. Thus robots that perform delivery tasks such as our ball delivery task either have specific hand-coded rules to regularly check if the ball has fallen out of the gripper or do not check at all.

In theory, one could imagine driving a conventional serial reasoner with perceptual input by interfacing the perceptual system to the symbolic database so that the outputs of the perceptual systems are something like first-order logic assertions expressed as tree-like data structures. It is difficult to imagine how to do this in a domain-independent manner, however, since the perception system needs to know what information is relevant to the current task. Another issue is how to inform the reasoner about changes in the environment. A planner that is deep in a non-deterministic search may need to completely restart the search when the environment changes. Determining whether a given change is relevant and requires a restart of the search process is not an inviting problem. Another possibility is to build a perceptual system that emulates the symbolic database (Horswill, 1995). While useful, this approach still leaves open the problem of tracking environmental changes and updating stale inferences.

By far, the simplest way to ground an inference process is to choose a representation language whose inference rules can be compiled into parallel networks and then drive the inputs of those networks with the outputs of the perceptual system. Then the outputs of the inference rules will follow the perceptual system as its state changes. The issue, then, is to find the most useful representation language for which we can do this.

2. Role passing

We have developed a set of techniques for implementing limited inference and variable binding in otherwise behavior-based architectures. We call the class of architectures they encourage "role passing" architectures for reasons that should be clear in a moment. Role passing allows the use of more expressive representation languages without sacrificing parallelizability or the ability to ground inference in sensor data.

A role-passing system is composed of four basic types of components:

- A set of traditional sensory-motor *behaviors*, as in subsumption (Brooks, 1986) or the Schema system (Arkin, 1997).
- A set of specialized *short-term memory* systems. Each STM provides a way of representing a specific type of information about objects in the world. Internally, an STM is composed of a fixed pool of representations that can be *allocated* to hold information about a given object. Allocated representations are *tagged* with a specific tag (see below) that is used when looking up information about the object being represented. Once allocated and tagged, the representation is said to be *bound* to that tag. The set of possible tags is fixed in advance and is assumed to be relatively small.
- A set of *access ports* between behaviors and STMs. When a behavior drives the input of an access port with a tag, the STM drives the port's output with the data from the representation bound to that tag.
- An *interference network* that receives data from the STMs, performs deductions using it, and generates control signals to drive the behaviors and STMs. This will be discussed more fully in section 3.

As with any parallel reactive system, these components should be thought of as pieces of hardware that continually recompute their outputs based on their current inputs, although in practice all our implementations run on standard serial computers.

Most of the communication within a role-passing system consists of passing tags from one system to another. When a behavior requires an object as a parameter, that object is specified by its tag. The behavior routes the tag to the appropriate access port and receives the data it needs about the object from the port's STM.

The term "short-term memory" is somewhat misleading, since it implies a passive information store. Our use here is broader, however. In particular, we will group attentive perceptual systems, such as visual object trackers, into STMs. Such a collection of trackers is a memory insofar as it remembers to what objects the system is attending, and to what tags those objects are bound. It is, however, an "active" memory in the sense that it continually updates its own representations.

Our current implementation (see section 4), uses three STMs:

- A *visual tracking system* that can search for and track the positions of a set of designated objects. The system is composed of a set of individual trackers (3, in our implementation). Other components in the system can request that an object be tracked by specifying a color to search for and a tag to bind it to. The tracking system will then allocate a tracker to it, bind the tracker to the specified tag, and set it searching for the specified color.

- An *odometric tracking system* that can dead-reckon the positions of places the robot has been before. Other components can request that places be remembered by providing a tag to which to bind the place representation.
- A *description buffer* that stores information about the appearances of different objects. The present implementation stores statistics about the colors on an object's surface (e.g. mean RGB value, variances, etc). The descriptions can be used as input to the tracking system to initiate a search for an object with the specified statistics. Descriptions are bound by providing a set of statistics and a tag value to which to bind them. Unlike the other two STMs, the description buffer really is a passive information store.

Each of these STMs can potentially hold information about several objects. A given object can be represented in any, all, or none, of the STMs at a given time. Although we will focus on these three examples, many other types of STMs are possible. For example, Matarić's active map representation (Matarić, 1992) could be considered an STM in our sense, as could marker-passing knowledge bases such as NETL (Fahlman, 1979).

The most speculative aspect of the architecture is the choice of tags. In role-passing systems, we have used linguistic role names (e.g. AGENT, PATIENT, SOURCE, DESTINATION) as tags. In other words, we assume that linguistic role names (and perhaps other features) are the keys by which short term memory is accessed. This may or may not be a good policy in the long run, but it has been useful so far. In any case, it is independent of the other architectural decisions, such as the use of tagging in general or the decision to break short-term memory up into specialized subsystems.

We can now sketch the solution of the delivery problem using role-passing. We bind the color of the object we want to deliver to the role PATIENT in the description buffer. We also bind our current location to the role DESTINATION in the odometric tracking system. We use an inference network that implements the following rules:

- Assert goal(in-hand(PATIENT))
- If in-hand(PATIENT),
  then assert goal(near(DEST))
- If in-hand(PATIENT) and near(DEST), then done
- If goal(in-hand(X)) and not(in-hand(X)), then assert goal(near(X))
- If goal(in-hand(X)) and not(in-hand(X)) and near(X), then activate the GRAB behavior with X as its argument
- If goal(near(X)),
  then assert goal(know(distance(X)))
  and goal(know(direction(X)))
- If goal(near(X)) and know(distance(X)) and know(direction(X)), then activate the DRIVETO behavior with the argument X.
- If goal(know(distance(X)))
  and not(know(distance(X)))
  and know(description(X)), then activate the visual search system with an argument of X.

and run the rules continuously and in parallel. At any given moment, the robot will drive to the destination iff current sensor data indicates that the ball is in the gripper. Similarly, it will attempt to acquire the ball iff sensor data indicates it is not in the gripper. Like parallel reactive systems, the robot will recover automatically from losing the ball. However, we can change the object it delivers or the destination it takes it to just by changing the initial role bindings. We do not have to duplicate rules or behaviors for each possible ball.

3. Grounding inference

Given the basic role-passing architecture, we can implement an inference network that supports limited forms of variable binding by treating roles as bound variables. Alternatively, we can think of roles as names with indexical reference. Since role names typically have functional significance (e.g. something bound to DESTINATION must be a place to which something can, has, or wants to go), role-passing can be thought of as a kind of indexical-functional (deictic) representation (Agre and Chapman, 1987, Agre, 1988).

3.1. *Representing functions, predicates, and individuals*

Peripheral systems, such as the behaviors and STMs are free to use whatever internal representations are convenient (images, sonar data, grid-based representations of space, etc.). However the inference network only uses three representations: short term memory keys (roles/tags), unary predicates, and scalar functions.

Predicates are represented by giving their values on every object bound in the system. Since there are only a finite number of roles, we can represent the value of a predicate as a bit-vector in which the ith bit being set means that the the predicate is true of the object bound to the ith role:

predicate	source	dest	patient	agent
near	T	F	F	T
see	F	T	T	F

As a slight abuse of terminology, we will call this the *extension* of the predicate. For convenience sake, we will use the same representation for roles when they are used as short-term memory keys: to refer to the object bound to the ith role, we use the bit-vector with only the ith bit set.

We can use a similar representation for real-valued functions over objects. A function is represented as a vector of numbers in which the ith value of the vector is the value of the function on the object bound to the ith role:

function	source	dest	patient	agent
distance	15	0	0	17
direction	14	87	35	-28

Since we can represent predicates and functions as small vectors, we can efficiently implement them on both von Neumann and parallel hardware. On von Neumann hardware, a function is a standard vector of memory cells and a predicate can be represented as a single memory word. Moreover, we can compute logical connectives with single machine instructions. A compound expression like:

$$P(x) \wedge (Q(x) \vee W(x))$$

can be evaluated by the code:

```
(logand p (logior q w))
```

(or `p&(q|v)` in C). The expression computes the bit-vector of the individuals for which the subexpression $P(x) \wedge (Q(x) \vee W(x))$ holds (assuming p, q, and w hold the bit-masks for P, Q, and W, respectively).

The representation is also conveniently implemented in a feed-forward network. Each predicate or function is implemented as a compact bus of wires holding the symbol's value for each indexical. Logical connectives are again implemented as bit-wise logic operations, but this time by constructing separate gates.

Thus, by limiting our representation to a small set of indexical names, and by limiting ourselves to single-place predicates and function symbols, we can efficiently represent logical expressions as fairly compact dependency networks. These networks are easily parallelized and easily updated in real-time.

3.2. *Representing propositional attitudes*

Thus far, we have no way of representing the distinction between $near(destination)$ being false, and $near(destination)$ being unknown. The distinction is critical for reasoning about attention.

The problem is easily solved by keeping a vector of valid bits in parallel with the truth bits of predicates and values of functions. Within the framework of logic, these valid bits are naturally interpreted as assertions about the agent's state of knowledge: if the truth bits encode the proposition $P(x)$, then the valid bits encode the proposition "know-whether $P(x)$", which we will write $\Delta(P(x))$:

$$\Delta(P(x)) \equiv K(P(x)) \vee K(\neg P(x))$$

$$\Delta(f(x)) \equiv \exists y.K(f(x) = y)$$

Since we ultimately want to build a problem solver, it is also useful to add a "goal" modality, so that we can express that $P(x)$ is a goal or even that $\Delta(P(x))$ is a goal. Thus the real representation of predicates is not just a bit-vector, but four bit-vectors: the predicate's extension, its valid bits, its goal bits, and its knowledge-goal bits. Functions are three bit-vectors and a scalar vector because functions are scalar-valued, not truth valued.

3.3. *Derived functions and predicates*

We can now define functions and predicates in terms of other functions and predicates. Given an axiom of the form

$$\forall x.P(x) \equiv \Phi(Q_1(x), ..., Q_n(x))$$

we can generate hardware or machine code to compute P from Q_i as described in section 3.1. We can also generate the axiom

$$\forall x.\Delta(Q_1(x)) \wedge ... \wedge \Delta(Q_n(x)) \Rightarrow \Delta(P(x))$$

which, again, is easily implemented as a bit-vector operation, this time on the valid bits.

At the cost of extra complexity, we can make this axiom stronger, at least for the standard logical connectives. For example, if we have that

$$\forall x.P(x) \equiv Q(x) \wedge W(x)$$

then we have that

$$\forall x.\Delta(P(x)) \equiv$$
$$(\Delta(Q(x)) \wedge \Delta(W(x))) \vee K(\neg Q(x)) \vee K(\neg W(x))$$

Goals are a more complicated manner. If we define $P(x)$ as $Q_1(x) \wedge ... \wedge Q_n(x)$, then we may often want to include the axioms:

$$\forall x.goal(P(x)) \Rightarrow goal(Q_1(x)) \wedge ... \wedge goal(Q_n(x))$$

$$\forall x.goal(\Delta(P(x)))$$
$$\Rightarrow goal(\Delta(Q_1(x))) \wedge ... \wedge goal(\Delta(Q_n(x)))$$

However, this can get us into trouble, since it may be important to solve one subgoal before the other. Thus while these axioms are useful defaults, they are not universally true. In our implementation, we allow conjunctions to solve their subgoals either in parallel, as above, or serially (see section 4).

3.4. *Grounding primitive percepts*

In order for our predicate/function representation to be useful, the perceptual systems must be able to generate it themselves. Fortunately, this is easy to do.

Recall that we have grouped perceptual systems together into STMs. Each STM contains a homogeneous pool of perceptual systems, such as visual object trackers or odometric trackers, or passive representations such as color descriptions. Each STM allocates perceptual systems (trackers, in this case) on demand and keeps track of the roles to which they are bound. The perceptual systems, for their part, measure various functions and predicates. For example, the visual object trackers might measure the predicate *is-moving* and the functions *distance* and *direction*. The STM can compute the complete extension of a predicate, such as *is-moving*, by forming a bit-vector of all the role tags of all the trackers that are reporting motion. It can compute the valid bits by forming a bit-vector of all the roles to which any tracker is bound.

In general, for an STM comprised of perceptual systems $T_1, ..., T_l$, each of which measures a set of predicates $P_1, ..., P_m$, and a set of functions $f_1, ..., f_n$), we can compute the extension and valid bits of P_i or f_i as:

$$\text{ext}(P_i) = \bigvee_{\{j|P(T_j)\}} \text{role}(T_j) \tag{1}$$

$$\text{ext}(\Delta(P_i)) = \bigvee_j \text{role}(T_j) \tag{2}$$

$$\text{f}_\text{i}(r) = \begin{cases} f(T_j), & r \in \text{role}(T_j) \\ \text{undefined}, & r \text{ not bound} \end{cases} \tag{3}$$

$$\text{ext}(\Delta(f_j)) = \bigvee_j \text{role}(T_j) \tag{4}$$

where, $P_i(T_j)$ is T_j's measurement of P_i, $f_i(T_j)$ is T_j's measurement of f_i, ext(P_i), P_i's measured extension, is the bit-vector of roles for which P_i's has been measured by some T_j to be true, ext($\Delta(P_i)$) is the bit-vector of roles for which P_i's value has been measured (be it true or false), and role(T_j) is the the role (in bit-vector format) to which T_j is bound. Passive STMs, such as the description buffer, report their data to the inference network in the same way.

3.5. *Specifying arguments to behaviors*

Behaviors, such as the DRIVETO behavior, take as arguments objects in the world. Ultimately, DRIVETO has to obtain direction and distance information about the object from either a visual tracker bound to the object or an odometric tracker bound to it. It gets this information through its access ports to the visual tracking STM and the odometric tracking STM. It drives the role input of the port with the bit-vector representing X. The access port compares it to the bindings of each tracker in the STM and drives the output of the port with the distance and direction signals of the matching tracker (see figure 4).[3] DRIVETO then uses whatever distance and direction data is presented on the output of the port to steer the robot appropriately.

Our delivery control system included the rule

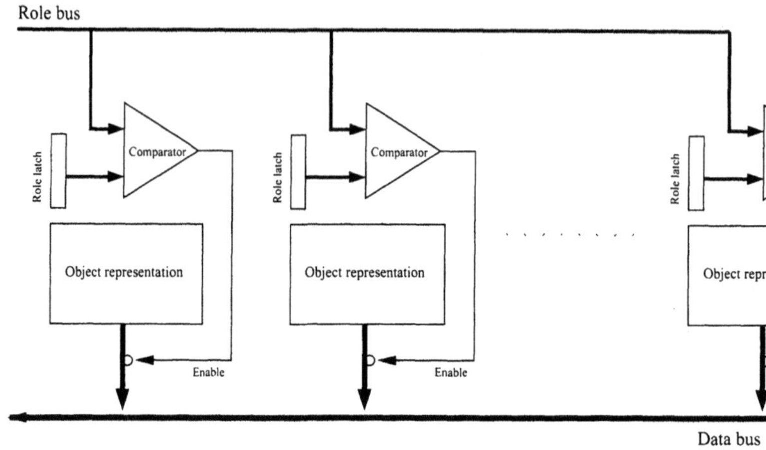

Fig. 2. Access ports. Each object representation (tracker, description, etc.) stores the set of roles to which it is bound in a role latch. When a behavior is activated, its argument is specified by driving the role bus with the argument's role. The tracker whose role latch matches the role bus then drives the output bus with its data.

- If goal(near(X)) $\wedge$ $\triangle$(distance(X)) $\wedge$ $\triangle$(direction(X)), then activate the DRIVETO behavior with the argument X.

which we can implement by computing the compound predicate D, given by

$$D(x) \equiv$$
$$goal(\text{near}(x)) \wedge \triangle\text{distance}(x) \wedge \triangle\text{direction}(x)$$

and routing its extension to the argument input of DRIVETO. Whenever D has a non-null extension and one of the trackers has a binding for one of the objects in its extension, then the tracker will drive the output of DRIVETO's access port with that object's direction and distance information, causing the robot to drive toward it.

An important question is how to resolve conflicts when D's extension has many objects. Thus far, we have dealt with the problem by writing the control rules so it can't happen.

3.6. *Equality*

Sometimes, it is useful to be able to assert that two roles co-refer. For example, equality assertions could be used to implement an alternative form of parameter passing. Rather than DRIVETO taking a tag as an input, DRIVETO could be designed to drive to any object bound to the role TARGET. DRIVETO could then be "called" on some other object, such as PATIENT, by asserting the equality TARGET=PATIENT.

In the general case, there are n^2 possible assertions of equality between n roles. The current set of equality assertions can be represented as an $n \times n$ matrix $\mathbf{E}$ in which E_{ij} is true iff i and j are asserted to be equal. Given the matrix, we can map a set of names $\mathbf{n}$ to the set of names, $\mathbf{n}_E$, equivalent to it, by performing a matrix multiplication:

$$\mathbf{n}_E = (\mathbf{I} + \mathbf{E})\mathbf{n}$$

Thus we can assert that two names n_i and n_j co-refer by asserting the entries E_{ij} and E_{ji} in the matrix.

In the general case, we must first compute the transitive closure of the asserted equalities. Transitive closure can be computed iteratively by repeated matrix multiplication. Alternatively, if we can place an ordering on the names such that equality assertions are only made between names adjacent in the ordering, then $\mathbf{E}$ will be in block diagonal form and we can compute the mapping directly as:

$$\mathbf{n}_E = (\mathbf{I} + \mathbf{E_1})(\mathbf{I} + \mathbf{E_2})...(\mathbf{I} + \mathbf{E_m})\mathbf{n}$$

where the $\mathbf{E}_1$ are the blocks of $\mathbf{E}$.

One of the advantages of this scheme is that the matrix multiplications need only be performed in the STMs. Each tracker (or other representation) multiplies the set of roles to which it is bound by $\mathbf{I} + \mathbf{E}$ to obtain the set of roles equivalent to the ones to which its bound, then uses those roles in driving the predicate and function busses. Thus, we modify eqs. 1 as follows:

$$\begin{aligned} \mathrm{ext}(P_i) &= \bigvee_{\{j|P(T_j)\}} (\mathbf{I}+\mathbf{E})\mathrm{role}(T_j) \\ \mathrm{ext}(\triangle(P_i)) &= \bigvee_j (\mathbf{I}+\mathbf{E})\mathrm{role}(T_j) \\ \mathrm{f_i}(r) &= \begin{cases} f(T_j), & r \in (\mathbf{I}+\mathbf{E})\mathrm{role}(T_j) \\ \text{undefined}, & r \text{ not bound} \end{cases} \\ \mathrm{ext}(\triangle(f_j)) &= \bigvee_j (\mathbf{I}+\mathbf{E})\mathrm{role}(T_j) \end{aligned}$$

The derived predicates and functions will then be closed under the current set of equality assertions. This lets us avoid computing equality separately at each predicate or function. For t trackers and n roles, the scheme then requires $O(tn^2)$ hardware, or, provided the roles fit in a single machine word, $O(tn)$ time on a von Neumann machine.

4. Implementation

We have implemented role-passing on a vision-based mobile robot called Kludge (see figure 1). Kludge is built from a shortened RWI B14 enclosure with an MIT/DIdeas Cheap Vision Machine serving as its main computer (see figure 1). The CVM is a 25MIP floating-point DSP with 1MB of RAM and a memory-mapped frame grabber. The CVM can perform a wide range of real-time vision operations and consumes only 5-10W. Kludge is also equipped with a pair of servo-controlled mandibles suitable for grasping a set of balls. Its visual and motor capabilities allow it to find and grab balls, bring them to people, or follow people, given knowledge of the color of their clothing. It is also used to play games with children.

Kludge implements the visual tracking STM, the odometric tracking STM, and the description buffer discussed above. Its control rules implement a range of fetching and following tasks, which we describe below. The rules are expressed in a language called *microdoer*, which is similar in spirit to MICROPLANNER or Prolog. The microdoer interpreter simulates the feed-forward network inference network described above and calls the appropriate routines for updating behaviors and perceptual routines. On every cycle of its main control loop, it acquires a new image, runs the vision system and other sensor systems, updates the sensory outputs, reevaluates all the nodes in the inference network, and, finally, runs the behaviors.

Table 1. Roles in the delivery task.

Role	Meaning
Patient	The ball to be delivered
Source	Starting location of the robot
Destination	Person or place to which to deliver the object

The microdoer interpreter itself runs under a simple Scheme interpreter written in C. Kludge's vision code is written in C and is called from Scheme. The vision system allows it to search for and track several objects simultaneously. Search and tracking are based on color. The vision system can also perform real-time collision avoidance using the Polly algorithm (Horswill, 1994). Since the Microdoer interpreter is running on a low performance interpreted lisp, it is much slower than it ought to be. The set of rules given here take approximately 50ms to update on each clock cycle. Since the vision system and the generation of debugging output also each take around 50ms, the frame rate of the system varies between 7Hz and 10Hz depending on the amount of debugging output being generated.

The various STMs compute 13 predicates and functions (see figure 3). Most are either *boundp* predicates (true iff a given STM binds a given role), *outputs* (real-valued functions), or *thresholds* (true if a given function is above or below a given threshold).

4.1. *Rules for fetch-and-follow tasks*

Given these sensory predicates, it is straight-forward to write rules to implement fetching, following, and delivering various objects. One rule tells the system that it should use the `driveto` behavior to establish the `near` predicate. Another rule tells it that to achieve the `inhand` predicate, it needs to be near the object and facing it, then open the mandibles and drive forward. A separate set of mandible control rules grab any object that is seen to be in the grabbing area.

Some of the control rules are implemented as conjunctions and disjunctions of sensory predicates (see figure 4). These are computed as in section 3.3. Conjunctions come in two forms depending on whether they spawn their subgoals serially or in parallel. Parallel conjunctions, when goals, always make their conjuncts goals, whereas serial conjunctions first make their first conjunct a goal, then, when it is achieved, make the second conjunct a goal, and so on. Subgoaling is updated on every clock cycle, so if the first conjunct becomes false, the other conjuncts lose their goal

```
(define-STM-boundp-node have-description description-STM #f)
(define-STM-boundp-node see tracker-STM have-description)
(define-STM-output-node distance tracker-distance tracker-STM see)
(define-STM-output-node image-x-coordinate tracker-x tracker-STM see)

(define-threshold-node visually-near < distance 13)
(define-threshold-node inhand < distance 3)
(define-test-node visually-facing image-x-coordinate centered?)

(define-STM-predicate odometrically-facing
  odometer-STM odometrically-facing? #f)
(define-STM-predicate odometrically-near
  odometer-STM odometrically-near? #f)
(define-STM-boundp-node odometrically-tracking odometer-STM #f)
```

Fig. 3. Sensory predicates and functions in Kludge.

status. Disjunctions do not perform goal regression at all.

In addition to conjunctions and disjunctions, there are a set of explicit goal reduction rules linking goals to behaviors or subgoals (see figure 5). The semantics of the rules are as follows:

- (when-unsatisfied P B): when $P(r)$ for some role r is an unsatisfied goal, fire the behavior $B(r)$.
- (reduce-unsatisfied P_1 P_2): when $P_1(r)$ is an unsatisfied goal, assert $goal(P_2(r))$.
- (do-all! P $proc$): run $proc$ on every role for which P is true.
- when: as in Common Lisp.

These control rules allow the system to perform a range of actions from simply driving up to a place or object to delivering a specified object to a specified place. The choice of the task is determined by the initial bindings of representations to roles and by the goals asserted in the system. To grab a red object, bind the red description to $patient$ and assert $goal(inhand(patient))$. To drive to it without grabbing it, bind it to $destination$ and assert $goal(near(destination))$ instead. To drive to a specified place, bind its odometric position to $destination$ instead of the red description.

4.2. *Following instructions*

To test out the system, we wrote a simple finite-state parser. While simple, it has been useful for debugging and experimentation.

The parser keeps the user's input in a shift-register and parses one word per control loop cycle. A separate shift register holds the role names that successive syntactic components of the current verb are to take. Encountering a verb asserts a goal and loads the role register. Encountering a noun binds the representation listed in its lexical entry to the current role. Encountering a preposition changes the current role. The conjunction "then" (as in "do A then do B") halts the parser until the current goal is achieved. There are also a few modifiers, such as "continually" which changes verbs from specifying goals of achievement to maintenance goals. As a final feature, the action rule `(when-unsatisfied boundp query-user)` (see figure 5) fires whenever one of the roles of the current verb has been left unspecified. It halts the robot and displays the message "what?" or "where?", depending on the role that was left unbound. If the user replies (i.e. with "home" or "to me"), the parser restarts and fills in the missing binding.

It must be stressed that this is an extremely simplistic model of parsing. It does not support determiners and it treats color words as nouns. It is limited to sentences like:

```
follow me
follow blue
```

```
(define-disjunction near (visually-near odometrically-near))
(define-disjunction facing (visually-facing odometrically-facing))
(define-disjunction boundp (see have-description odometrically-tracking))

(define-parallel-conjunction grabable (near facing))
(define-serial-conjunction grab (grabable mandibles-open approach))
```

Fig. 4. Other compound predicates in Kludge.

Table 2. Kludge's lexicon

Item	Category	Roles	Action
bring	verb	*patient destination*	assert $goal(inhand(patient))$, then assert $goal(near(destination))$
get	verb	*patient*	assert $goal(inhand(patient))$
go	verb		assert $goal(near(destination))$
come	verb	*destination*	assert $goal(near(destination))$
follow	verb	*patient*	assert $goal(near(patient))$
face	verb	*patient*	assert $goal(facing(patient))$
wait	verb		wait until time in time-buffer
turn	verb		turn number of degrees in turn-buffer
then	conjunction		halt parser until current goal is satisfied
home	noun		bind current location in odometry STM to current role
kludge	noun		bind kludge description to current role
blue	noun		bind blue description to current role
green	noun		bind green description to current role
red	noun		bind red description to current role
pink	noun		bind pink description to current role
black	noun		bind black description to current role
me	noun		bind black description to current role
to	preposition	*destination*	set current role to *destination*
from	preposition	*source*	set current role to *source*
continually	modifier		set don't-terminate flag
seconds	modifier		set time buffer to current time plus number buffer
degrees	modifier		set turn buffer to number buffer
left	modifier		negate turn buffer
right	modifier		none

```
get blue
bring blue here
bring blue to me
bring blue to red
go to blue then wait 10 seconds then
   bring blue to me
```

The system is not meant as a model of human comprehension.

4.3. *Example*

Table 3 gives an example trace of the system's behavior for the command "bring red to me". Note that unexpected contingencies, such as the human stealing the ball from it or its target straying out of view, do not cause execution failures — they simply cause it to rerun earlier actions (c.f. Nilsson (Nilsson, 1994)).

5. Discussion

Role-passing is an alternative hybrid control architecture. Unlike previous attempts to extend behavior-based systems (Rosenschein and Kaelbling, 1986, Maes, 1989), it allows limited predicate/argument structure in reasoning and passing of parameters to behaviors. This improves modularity and reduces computational complexity. It allows a single control system to implement "go to the red ball," "go to the blue ball," and "go to odometric location (x,y)" without having separate behaviors for each possible destination. It also makes it practical to write at least a simplistic NL parser.

```
;; Performing grab will make an object be in the hand.
(reduce-unsatisfied inhand grab)

;; Run the driveto behavior to get near an object
(when-unsatisfied near driveto)

;; Run the visual-search! behavior when you need to see an object.
(do-all! (logand (unsatisfied-goal see)
 (know-that have-description))
 visual-search!)

;; Control mandibles
(when-unsatisfied mandibles-open open-mandibles!)
(when (not-exists? (know-that near))
  (fold-mandibles!))
(when (exists? (know-that inhand))
  (close-mandibles!))

;; Query the user if they didn't specify a desired role.
(when-unsatisfied boundp query-user)
```

Fig. 5. Explicit action and reduction rules.

Unlike tiered architectures (Hexmoor, Horswill and Kortenkamp, 1997) which rely on conventional symbolic databases, role-passing has a tractable epistemic model. Behaviors, perceptual operations, and the reasoning system all communicate using a standard mechanism in which objects are referred to by their role tags rather than through lisp symbols (BEE0001) or S-expressions encoding definite descriptions (Firby, 1989). The bit-vectors used in role passing are easy for the perceptual system to generate and for the behaviors to decode. This makes it practical to completely update the world model on every cycle of the control loop. Since the inference rules can be compiled into a dependency network, the robot can also update all its inferences on every cycle, allowing the system to detect and respond to execution errors or other contingencies immediately.

By representing the agent's goals and state of knowledge explicitly using modals, the system can automatically allocate perceptual attention by regressing knowledge goals through logical connectives to determine the epistemic actions needed to achieve them. This simplifies the design of the system and relieves the programmer of the need to salt her axiomatization with separate epistemic rules stating when and how to check preconditions and postconditions.

The cost of these benefits is reduced expressiveness. Role-passing cannot handle term expressions, binary predicates or functions with arbitrary image sets. It is too weak to express the situation calculus. While some limitations must be accepted in order to assure real-time performance, there may be better trade-offs that can be made.

Certain features, however, appear very difficult to support in any such system. Term expressions require the dynamic allocation and mutation of arbitrary tree-structures, which are extremely difficult to parallelize. This amounts to solving the connectionist *binding problem* (Smolensky, 1990). So far as is known, any general solution to implementing arbitrary dynamic tree and pointer structures will either require a large (potentially $O(n^2)$) switching network to route signals, or will impose serial bottlenecks (effectively reducing it to a von Neumann machine). The fact that matching term expressions (unification) is P-complete (Dwork, Kanellakis and Mitchell, 1984) and is therefore believed to be unparallelizable is not encouraging.[4]

Table 3. Example execution trace for the command "bring red to me." Red is interpreted as any red object, and me is interpreted as any sufficiently large black object (the human is assumed to be wearing black jeans).

Event or state	Response
$goal(inhand(patient))$	Assert $goal(see(patient))$
$have - description(patient)$	Fire visual-search!$(patient)$
$see(patient)$	Fire $driveto(patient)$
$near(patient) \wedge facing(patient)$	Fire open-mandibles and approach
Patient stolen by human	Close mandibles, abort approach, re-fire visual-search
$see(patient)$	Fire $driveto(patient)$
$near(patient) \wedge facing(patient)$	Fire open-mandibles and approach
$inhand(patient)$	Fire close-mandibles, assert $goal(near(destination))$
$goal(near(destination))$	Assert $goal(K(distance(destination)))$
$goal(K(distance(destination)))$	Assert $goal(see(destination))$
$have - description(destination)$	Fire visual-search!$(destination)$
$see(destination)$	Fire $driveto(destination)$
Dest goes out of view	Abort driveto, fire visual-search!$(destination)$
$see(destination)$	Fire $driveto(destination)$
$near(destination) \wedge facing(destination)$	Fire open-mandibles

While role passing doesn't support term expressions, it does allow the binding of a small, fixed set of top-level variables, an approach which is gaining popularity. Minsky has proposed implementing communication within systems using global busses with task-specific semantics ("c-lines" (Minsky, 1977), and later "pronomes" (Minsky, 1986)). Shastri and Ajjanagadde describe a solution to the binding problem that supports a small number of general variables (Shastri and Ajjanagadde, 1993). (Indeed, Shastri uses bit-vector representations for implementing his scheme on von Neumann architectures.) (Agre and Chapman, 1987, Agre 1988), discuss the use of indexicals bound in the perceptual system as a replacement for variables bound by switching networks in the central system.

Like Agre and Chapman's deictic representation, role passing requires that the indexical names have functional significance; that they are effectively roles (slots) of what would otherwise be represented in some kind of frame structure. However, Agre and Chapman use extremely specific roles names. Microdoer requires that there be a relatively small number of role names that get recycled from situation to situation. Whereas Agre and Chapman's Pengi system has indexicals like `the-bee-that-is-chasing-me`, a role passing system would have to use a more generic role name like $threat$.

One advantage of giving indexicals specific functional significance (rather than making them more like normal variables) is that it gives us some leverage in representing higher-order predicates. Microdoer cannot represent a relation like $deliver(A, B)$ because its extension would require a quadratic number of bits. By analyzing deliver as

$$deliver \wedge patient = A \wedge destination = B$$

rather than as a two-argument relation, we can represent the concept of delivery. There is still a cost, however: we cannot represent two different instances of delivery at once because they would require inconsistent bindings of $patient$ and $destination$. Thus, the technique of modeling binary predicates with indexicals is much less expressive than full first-order logic. It cannot, for example, express the situation calculus. However, it buys a considerable complexity improvement over languages that can (e.g. Levesque et al., 1996) and also provides us with a simple epistemology. If the higher arity predicates that arise in practice can't be handled by some simple mechanism like indexicals, then it is unclear how we can ground them efficiently in real time at all.

As mobile robots and active vision systems become more capable, the need for powerful, expressive control languages will grow. A promising approach is to look for maximally expressive languages that are efficiently mappable into dependency networks. Parallel networks give good performance, but, more importantly, can be grounded efficiently in sensors. Role-passing is one example of how this can be done, and microdoer is one example of how a programming language can be built around role-passing. Better architectures and languages will no doubt be be developed as the robotics community attempts increasingly complex tasks.

Acknowledgements

Many thanks to Leslie Kaelbling, Lars Bergstrom, and Ivan Yen, for providing useful comments on this paper. Support for this research was provided in part by the National Science Foundation under grant number IRI–9625041. The Institute for the Learning Sciences was established in 1989 with the support of Anderson Consulting, part of the Arthur Anderson Worldwide Organization.

Notes

1. Real robots almost never use a purely reactive or deliberative system in the sense used here. For example, the subsumption architecture (Brooks and Connell, 1986) is not a reactive system. In fact, it has been used to build planners (Matariċ, 1992, Maes, 1989). However, they are still useful idealizations and so the debate within the community has been largely cast in terms of them.
2. In reality, the complexity of planning depends on many factors, see (Erol, Nau and Subrahmanian, 1995) for a survey of results on complexity bounds for different forms of planning.
3. This is how it would work in a real hardware implementation. In our serial implementation, behaviors are just subroutines that are called several times a second. Ports are implemented with a subroutine, called by the behavior, that scans the bindings of an STM and returns the output data for the tracker bound to the specified role.
4. Care must be taken in this argument, however. Any problem can be parallelized for bounded size inputs by using an exponential amount of hardware. Technically, the P-completeness result suggests that unification cannot be computed by small, shallow, feed-forward networks.

References

Agre, P.E. 1988. The dynamic structure of everyday life. Technical Report 1085, Massachusetts Institute of Technology, Artificial Intelligence Lab.

Agre, P.E. and Chapman, D. 1987. Pėngi: An implementation of a theory of activity. In *Proceedıngs of the Sixth National Conference on Artificial Intelligence*, pp. 268–272.

Arkin, R. 1997. *Behavior-Based Robotics*. MIT Press:Cambridge, MA.

Blake, A. and Yuille, A., editors. 1992. *Active Vision*. MIT Press: Cambridge, MA.

Bonasso, R.P., Firby, J., Gat, E., Kortenkamp, D., Miller, D.P., and Slack, M.G. 1997. Experiences with an architecture for intelligent reactive agents. In H. Hexmoor, I. Horswill, and D. Kortenkamp (eds.), *Special issue on software architectures for physical agents, Journal of Theoretical and Experimental AI*. Cambridge University Press.

Brooks, R.A. 1986. A robust layered control system for a mobile robot. *IEEE Journal of Robotics and Automoation*, 2(1):14–23.

Brooks, R.A. and Connell, J.H. 1986. Asynchronous distributed control system for a mobile robot. In *Cambridge Symposium on Optical and Optoelectronic Engineering*, SPIE.

Brown, C., Coombs, D. and Soong, J. 1992. Real-time smooth pursuit tracking. In A. Blake and A. Yuille, editors, *Active Vision*. MIT Press: Cambridge, MA., pp. 126–136.

Crisman, J.D. 1992. Color region tracking for vehicle guidance. In A. Blake and A. Yuille, editors, *Active Vision*. MIT Press: Cambridge, MA., chapter 7.

C. Dwork, C., Kanellakis, P., and Mitchell, J.C. 1984. On the sequential nature of unification. *Journal of Logic Programming*, 1(1):35–50.

Erol, K., Nau, D.S. and Subrahmanian, V.S. 1995. Complexity, decidability, and undecidability results for domain-independent planning. *Artificial Intelligence*, 76(1–2):75–88.

Fahlman, S.E. 1979. *NETL: A System for Representing and Using Real-World Knowledge*. MIT Press: Cambridge, MA.

Fairley, S.M., Reid, I.D., and Murray, D.W. 1995. Transfer of fixation for an active stereo platform vis affine structure recovery. In *Proceedings of the Fifth International Conference on Computer Vision*, pp. 1100–1105.

Firby, R.J. 1989. Adaptive execution in complex dynamic worlds. YALEU/CSD/RR 672, Computer Science Department, Yale University.

Hager, G.D. 1995. Calibration-free visual control using projective invariance. In *Proceedings of the Fifth International Conference on Computer Vision*, pp. 1009–1015.

Hasegawa, T., Nakano, Y.I., and Kato, T. 1997. A collaborative dialog model based on interaction between reactivity and deliberation. In W. L. Johnson, editor, *Proceedings of the First International Conference on Autonomous Agents*, Marina del Rey, CA USA, ACM SIGART, ACM Press, pp. 83–87.

Hexmoor, H., Horswill, I., and Kortenkamp, D. 1997. Software architectures for physical agents. In H. Hexmoor, I. Horswill and D. Kortenkamp, editors, *Special issue on software architectures for physical agents, Journal of Theoretical and Experimental AI*, Cambridge University Press.

Hexmoor, H., Horswill, I., and Kortenkamp, D., editors. 1997. *Special issue on software architectures for physical agents, Journal of Theoretical and Experimental AI*, Cambridge University Press.

Horswill, I. 1994. Collision avoidance by segmentation. In *Proceedings of the 1994 IEEE/RSJ Internation Conference on Intelligent Robots and Systems*, Munich, Germany, IEEE Press.

Horswill, H. 1995. In *Proceedings of the 14th International Joint Conference on Artificial Intelligence*, Montreal.

Huttenlocher, D.P., Noh, J.J. and Rucklidge, W.J. 1992. Tracking non-rigid objects in complex scenes. TR 93-1320, Computer Science Department, Cornell University.

Inoue, H. 1993. Vision based robot behavior: Tools and testbeds for real world ai research. In *Proceedings of the Thirteenth International Joint Conference on Artificial Intelligence*, Chambery, France, pp. 767–773.

Kaelbling, L.P. 1988. Goals as parallel program spcifications. In *Proceedings, AAAI-88*, St. Paul, MN, pp. 60–65.

Levesque, H.J. and Brachman, R.J. 1985. A fundamental tradeoff in knowledge representation and reasoning (revised edition). In R.J. Brachman and H.J. Levesque, editors, *Readings in Knowledge Representation*, Morgan Kaufman: Los Altos, CA, pp. 42–70.

Levesque, H.J., Reiter, R., Lespérance, Y., Lin, F. and Scher, R.B. 1996. Golog: A logic programming language for dynamic domains. *Journal of Logic Programming*.

Lowe, D. 1992. Robust model-based motion tracking through the integration of search and estimation. *International Journal of Computer Vision*, 8(2):113–122.
Maes, P. 1989. How to do the right thing. AI Memo 1180, MIT Artificial Intelligence Laboratory.
Matarić, M. 1997. Behavior-based control: Examples from navigation, learning, and group behavior. In H. Hexmoor, I. Horswill, and D. Kortenkamp (eds.), *Special issue on software architectures for physical agents, Journal of Theoretical and Experimental AI.* Cambridge University Press.
Matarić, M.J. 1992. Minimizing complexity in controlling a collection of mobile robots. In *IEEE International Conference on Robotics and Automation*, Nice, France, pp. 830–835.
Minsky, M. 1977. Plain talk on neurodevelopmental epistemology. In *Proceedings of the Fifth International Joint Conference on Artificial Intelligence*, Cambridge, MA, pp. 1083–1092.
Minsky, M. 1986. *The Society of Mind.* Simon and Schuster: New York, NY.
Nilsson, N.J. 1994. Teleo-reactive programs for agent control. *Journal of Artificial Intelligence Research.*
Prokopowicz, P.N., Swain, M.J. and Kahn, R.R. 1994. Task and environment-sensitive tracking. In W. Martin, editor, *Proceedings of the IAPR/IEEE Workshop on Visual Behaviors*, Seattle, pp. 73–78.
Rosenschein, S.J. and Kaelbling, L.P. 1986. The synthesis of machines with provable epistemic properties. In J. Halpern, editor, *Proc. Conf. on Theoretical Aspects of Reasoning about Knowledge*, Morgan Kaufmann, pp. 83–98 .
Shastri, L. and Ajjanagadde, V. 1993 From simple associations to systematic reasoning: A connectionist representation of rules, variables, and dynamic bindings using temporal synchrony. *Behavioral and Brain Sciences*, 16.
Smolensky, P. 1990. Tensor product variable binding and the representation of symbolic structures in connectionist systems. *Artificial Intelligence*, 46(1–2):159–216.
Ullman, J.D. and Wisdom, J. 1997. *A First Course in Database Systems*. Prentice-Hall.

Ian Horswill is an Assistant Professor of Computer Science at Northwestern University. He received his Ph.D. in Computer Science from the Massachusetts Institute of Technology. His research interests include planning, problem solving, vision, and robotics. However, his principle research focus is on integrated systems that cleanly link perception, reasoning, and action.

Autonomous Robots 5, 79–95 (1998)

Interleaving Planning and Robot Execution for Asynchronous User Requests

KAREN ZITA HAIGH AND MANUELA M. VELOSO
Computer Science Department, Carnegie Mellon University, Pittsburgh PA 15213-3891
http://www.cs.cmu.edu/~khaigh, http://www.cs.cmu.edu/~mmv
khaigh@cs.cmu.edu
mmv@cs.cmu.edu

Abstract. ROGUE is an architecture built on a real robot which provides algorithms for the integration of high-level planning, low-level robotic execution, and learning. ROGUE addresses successfully several of the challenges of a dynamic office gopher environment. This article presents the techniques for the integration of planning and execution.

ROGUE uses and extends a classical planning algorithm to create plans for multiple interacting goals introduced by asynchronous user requests. ROGUE translates the planner's actions to robot execution actions and monitors real world execution. ROGUE is currently implemented using the PRODIGY4.0 planner and the Xavier robot. This article describes how plans are created for multiple asynchronous goals, and how task priority and compatibility information are used to achieve appropriate efficient execution. We describe how ROGUE communicates with the planner and the robot to interleave planning with execution so that the planner can replan for failed actions, identify the actual outcome of an action with multiple possible outcomes, and take opportunities from changes in the environment.

ROGUE represents a successful integration of a classical artificial intelligence planner with a real mobile robot.

Keywords: continuous planning, robotic agents, execution monitoring, replanning

1. Introduction

We have been working towards the goal of building autonomous robotic agents that are capable of planning and executing high-level tasks, and learning from the analysis of execution experience. This article presents our work extending the high-level reasoning capabilities of a real robot. One of the goals of this research project is to have the robot move autonomously in an office building reliably performing office tasks such as picking up and delivering mail and computer printouts, returning and picking up library books, and carrying recycling cans to the appropriate containers.

*A preliminary version of this paper was presented at the First International Conference on Autonomous Agents in February 1997.

Our framework consists of integrating Xavier (O'Sullivan et al., 1997; Simmons et al., 1997) with the PRODIGY4.0 planning and learning system (Veloso et al., 1995). The resulting architecture and algorithms is ROGUE (Haigh and Veloso, 1997). ROGUE provides a setup where users can post tasks for which the planner generates appropriate plans, delivers them to the robot, monitors their execution, and learns from feedback from execution performance. ROGUE effectively enables the communication between Xavier, PRODIGY4.0 and the users.

In this article, we focus on presenting the techniques underlying the planning and execution control in ROGUE. The learning algorithm is under development and results are being compiled and can be found in (Haigh and Veloso, 1998). The planning and execution capabilities of ROGUE form the foundation for a complete, learning, autonomous agent.

ROGUE generates and executes plans for multiple interacting goals which arrive asynchronously and whose task structure is not known *a priori*. ROGUE interleaves tasks and reasons about task priority and task compatibility. ROGUE enables the communication between the planner and the robot, allowing the system to successfully interleave planning and execution to detect successes or failures and to respond to them. ROGUE controls the execution of a real robot to accomplish tasks in the real world.

1.1. System Architecture

Figure 1 shows the general architecture of the system. ROGUE accepts tasks posted by users, calls the task planner, PRODIGY4.0, which generates appropriate plans, and then posts actions to the robot, Xavier, for execution. ROGUE provides appropriate search control knowledge to the planner and monitors the outcome of execution.

Xavier is a mobile robot being developed at Carnegie Mellon University (O'Sullivan et al., 1997; Simmons et al., 1997) (see Fig. 2).

It is built on an RWI B24 base and includes bump sensors, a laser range finder, sonars, a color camera and a speech board. The software controlling Xavier includes both reactive and deliberative behaviours, integrated using the Task Control Architecture (TCA) (Simmons, 1994). TCA provides facilities for scheduling and

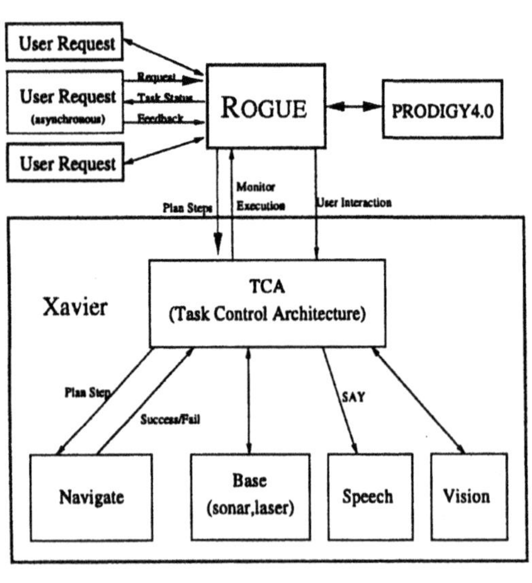

Figure 1. ROGUE Architecture.

Figure 2. Xavier the robot.

synchronizing tasks, resource allocation, environment monitoring and exception handling. The reactive behaviours enable the robot to handle real-time local navigation, obstacle avoidance, and emergency situations (such as detecting a bump). The deliberative behaviours include vision interpretation, maintenance of occupancy grids and topological maps, and path planning and global navigation. The underlying architecture is described in more detail by Simmons et al. (1997).

PRODIGY is a domain-independent planner that serves as a testbed for machine learning research (Carbonell et al., 1990; Veloso et al., 1995). The current version, PRODIGY4.0, is a nonlinear planner that follows a state-space search guided by means-ends analysis and backward chaining. It reasons about multiple goals and multiple alternative operators to achieve the goals. It reasons about interacting goals, exploiting common subgoals and addressing issues of resource contention.

Xavier reliably performs actions requested of it, but has no task planning abilities. PRODIGY4.0, meanwhile, is a complex task planner that had never been used in a real execution domain; as such, it had never been used for asynchronous goals or in an environment where the state spontaneously changes. In combining the two systems, the challenges for ROGUE include developing a communication mechanism for control and feedback, as well as extending the planner to handle the dynamics of a real-world task.

1.2. Planning and Execution in ROGUE

There are a few approaches to creating plans for execution. Shakey (Nilsson, 1984) was the first system to use a planning system on a robot. This project was based on a classical planner that ignored real world uncertainty (Fikes et al., 1972) and followed a deterministic model to generate a single executable plan. When execution failures occurred, replanning was invoked.

This pioneering approach has been acknowledged as partly successful, but also has been criticized for its lack of reactivity, and has led to significant research into planning systems that can handle the uncertainty of the real world. *Conditional planning* is one approach that aims at considering in the domain model all the possible contingencies of the world and planning ahead for each individual one (Atkins et al., 1996; Mansell, 1993; Pryor, 1994; Schoppers, 1989). In most complex environments, the large number of possible contingencies means that complete conditional planning becomes infeasible, but may nevertheless be appropriate in particularly dangerous domains.

Probabilistic planning takes a more moderate approach in that it only creates conditional plans for the most likely problems (Blythe, 1994; Dean and Boddy, 1988; Gervasio and DeJong, 1991; Kushmerick et al., 1993). It relies on replanning when unpredictable or rare events take place. Although this approach generates fast responses to most contingencies, it may miss potential opportunities that arise from changes in the world. It should be noted that none of these systems have ever been applied to a real robotic system.

Another moderate approach is that of *parallel planning and execution*, in which the planner and the executor are decoupled (Drummond et al., 1993; Lyons and Hendriks, 1992; McDermott, 1992; Pell et al., 1997). The executor can react to the environment without a plan. The planner continually modifies the behaviour of the executor to increase the goal satisfaction probability. This approach leads to a system with fast reactions, but a set of default plans needs to be preprepared, and in some situations may lead away from the desired goal. Furthermore, the planner creates its plans based on assumptions about the world that may have changed during planning time.

We take a third approach: that of *interleaving planning and execution*, as do several other researchers (Ambros-Ingerson and Steel, 1988; Dean et al., 1990; Georgeff and Ingrand, 1989; Nourbakhsh, 1997). Interleaving planning with execution allows the system to reduce its planning effort by pruning alternative possible outcomes immediately, and also to respond quickly and effectively to changes in the environment. For example, the system can notice limited resources such as battery power, or notice external events like doors opening and closing. In these ways, interleaving planning with execution can create opportunities for the system while reducing the planning effort.

One of the main issues raised by interleaved planning and execution is when to stop planning and start executing. Dean (1990) selects between alternative actions by selecting the one with the highest degree of information gain, but is therefore limited to reversible domains. Nourbakhsh (1997), on the other hand, executes actions that prefix all branches of a conditional plan created after making simplifying assumptions about the world. The assumptions are built so that the planner always preserves goal reachability, even in an irreversible world.

ROGUE has two methods for selecting when to take an action. The first method selects an action when it is the first in a chain of actions that are known to lead towards the goal. PRODIGY4.0 uses means-ends analysis to build plans backwards, working from the goal towards the initial state. Each action is described in terms of required preconditions and possible effects; actions are added to the plan when their effects are desirable. When all the preconditions of an action are believed to be true in the current state, ROGUE executes the action. Since PRODIGY4.0 already has a partial plan from the initial state to the goal state, the action ROGUE selects is clearly relevant to achieving the goal. Actions whose failures may lead to irreversible states are avoided until it has exhausted all other possible ways of reaching the goal. The second method is used when there are multiple actions available for selection. ROGUE selects between these actions in order to maximize overall expected execution efficiency.

When ROGUE selects an action for execution, it executes a procedure that confirms the preconditions of the action, then executes the action, and finally confirms the effects. In addition to the explicit confirmation of preconditions and effects of actions, our system also monitors events that may affect pending goals. Each goal type has a set of associated monitors that are invoked when a goal of that type enters the system. These monitors run parallel to planning and may modify the planner's knowledge at any time. A given monitor may, for example, monitor battery power or examine camera images for particular objects.

The ability to handle asynchronous goals is a basic requirement of a system executing in the real world. A system that only handles asynchronous goals in a first-come-first-served manner is inefficient and loses many opportunities for combined execution. ROGUE easily incorporates asynchronous goals into its system without losing context of existing tasks, allowing it to take advantage of opportunities as they arise. By intelligent combining of compatible tasks, ROGUE can respond quickly and efficiently to user requests.

Amongst the other interleaving planners, only PRS (Georgeff and Ingrand, 1989) handles multiple asynchronous goals. ROGUE however abstracts much of the lower level details that PRS explicitly reasons about, meaning that ROGUE can be seen as more reliable and efficient because system functionality is suitably partitioned (Pell et al., 1997; Simmons et al., 1997). NMRA (Pell et al., 1997) and 3_T (Bonasso and Kortenkamp, 1996) both function in domains with many asynchronous goals, but both planners respond to new goals and serious action failures by abandoning existing planning and restarting the planner. As stated by Pell et al., establishing standby modes prior to invoking the planner is "a costly activity, as it causes (the system) to interrupt the ongoing planned activities and lose important opportunities". Throwing out all existing planning and starting over not only delays execution and but also can place high demands on sensing to determine current status of partially executed tasks.

In summary, ROGUE's interleaving of planning and execution can be outlined in the following procedure for accomplishing a set of tasks:

1. A user submits a request, and ROGUE adds the task information to PRODIGY4.0's state.
2. PRODIGY4.0 creates a plan to achieve the goal(s), constrained by ROGUE's priority and compatibility knowledge, incorporating any new requests.
3. ROGUE sends selected actions to the robot for execution, confirming that its preconditions are valid.
4. ROGUE confirms the outcome of each action. If failure, ROGUE notifies PRODIGY4.0 and forces replanning.

In this section we have motivated our work and summarized the most relevant related work. In Section 2, we describe PRODIGY4.0, present how it plans for multiple asynchronous goals, and introduce ROGUE's priority and compatibility rules (steps 1 and 2 above). We include a detailed example of the system's behaviour for a simple two-goal problem, when the goals arrive asynchronously. In Section 3 we present execution, monitoring and how the system detects, processes and responds to failure (steps 3 and 4). Finally we provide a summary of ROGUE's capabilities in Section 4.

2. Planning for Asynchronous Requests

The office delivery domain involves multiple users and multiple tasks. A planner functioning in this domain needs to respond efficiently to task requests, as they arrive asynchronously. One common method for handling these multiple goal requests is simply to process them in a first-come-first-served manner; however, this method leads to inefficiencies and lost opportunities for combined execution of compatible tasks (Goodwin and Simmons, 1992).

ROGUE is able to process incoming asynchronous goal requests, to prioritize them, and to suspend lower priority actions when necessary. It successfully interleaves compatible requests and creates efficient plans for completing all the tasks.

2.1. *Receiving a Request*

Users submit their task requests through one of three different interfaces: the World Wide Web (Simmons et al., 1997), Zephyr (DellaFera et al., 1988; Simmons et al., 1997), or a specially designed graphical user interface (Fig. 3) (Haigh and Veloso, 1996).

The slots in this last interface are automatically filled in with default information available from the user's plan file, and the deadline time defaults to one hour in the future. The interface can be extended with additional tasks at any time.

Each of these three interfaces forwards the request to ROGUE by TCA messages. When each new request comes in, ROGUE adds it to PRODIGY4.0's list of unsolved goals, and updates the task model, as shown in Table 1. The literal `(needs-item <user> <item>)` indicates that a request, sent by user `<user>`, is pending. This function is domain-dependent because the literals added relate strictly to this domain; however,

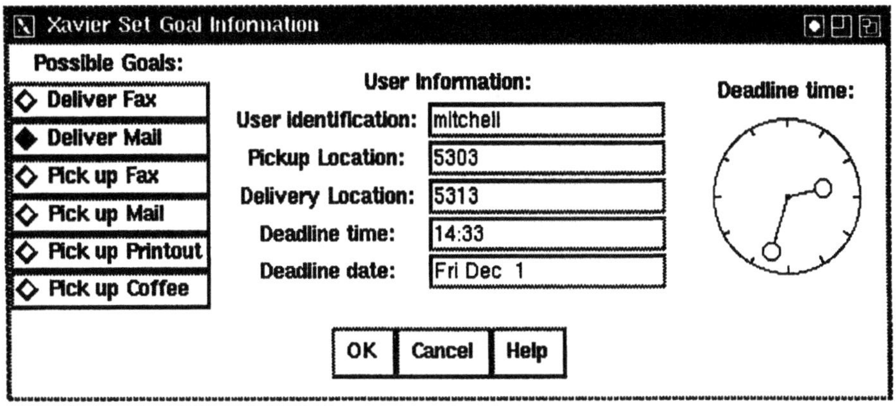

Figure 3. User request interface.

the structure would be identical for any other domain with asynchronous goals.

There is currently no explicit mechanism for a user to rescind a request; however PRODIGY4.0 will no longer plan for (or attempt to apply operators for) the associated top-level goal if it is simply removed from G and PG.

2.2. *Creating the Plan*

PRODIGY4.0 creates a plan for its unsolved goals by selecting operators whose effects achieve those goals. It continues adding operators to the incomplete plan until a solution to the problem is found. In Fig. 4 we show a simple incomplete plan.

The plan is built by a backward-chaining algorithm, which starts from the list of goals, G, and adds operators, one by one, to achieve its *pending goals*, i.e., to achieve preconditions of other operators that are not satisfied in the current state.

When all the preconditions of an operator are satisfied in the current state, PRODIGY4.0 simulates the effects of the action by *applying* the operator. Each time an operator is applied, the current state is updated with the effects of the action. PRODIGY4.0 terminates planning when each of the goals are true in the state.

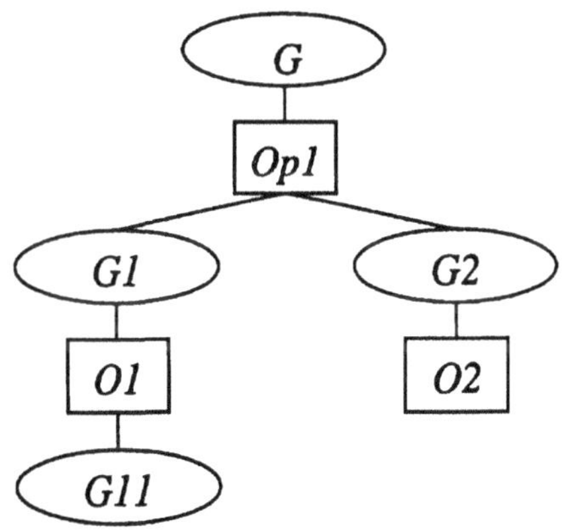

Figure 4. Representation of a plan. G is the top-level goal, and $Op1$ is the operator that achieves it. $G1$ and $G2$ are two preconditions of $Op1$ that are not true in the current state, and are achieved by $O1$ and $O2$ respectively.

Table 1. Integrating new task requests into PRODIGY4.0.

Define: $C \leftarrow$ current state
Define: $G \leftarrow$ top-level goal
Define: $PG \leftarrow$ pending goals cache (unsolved top-level goals and their subgoals)

Let R be the list of pending unprocessed requests
For each *request* $\in R$, turn *request* to goal:
- $C \Leftarrow C \cup \{$ (needs-item *request*-userid *request*-object)
  (pickup-loc *request*-userid *request*-pickup-loc)
  (deliver-loc *request*-userid *request*-deliver-loc) $\}$
- $G \Leftarrow$ (and G (has-item *request*-userid *request*-object))
- $PG \Leftarrow$ (and PG (has-item *request*-userid *request*-object))
- *request*-completed $\Leftarrow$ nil

Table 2. PRODIGY4.0 decision points, adapted from (Veloso et al., 1995).

PRODIGY4.0

1. If the goal statement *G* is satisfied in the current state, terminate.
2. Either (A) Subgoal: add an operator to *Plan* (**Back-Chainer**), or
   (B) Apply: apply an operator from *Plan* (**Operator-Application**).
   Decision point: Decide whether to apply or to subgoal.
3. Recursively call **PRODIGY4.0** on the resulting plan.

Back-Chainer

1. Pick an unachieved goal or precondition *g*.
   Decision point: Choose an unachieved goal.
2. Pick an operator *op* that achieves *g*.
   Decision point: Choose an operator that achieves this goal.
3. Add *op* to *Plan*.
4. Instantiate the free variables of *op*.
   Decision point: Choose an instantiation for the variables of the operator.

Operator-Application

1. Pick an operator *op* in *Plan* which is an *applicable operator*, that is the preconditions of *op* are satisfied in the current state.
   Decision point: Choose an operator to apply.
2. Update the current state with the effects of *op*.

The planning cycle involves several decision points, including which goal to select from the set of pending goals, and which applicable operator to apply. Table 2 shows the decisions made while creating the plans. **Back-Chainer** shows the decisions made while backchaining on the plan, **Operator-Application** shows how an operator is applied, and **Prodigy4.0** shows the top-level procedure.

Planning involves specifying a task model including operators, search control rules and domain description.

2.2.1. Operators. ROGUE's operators rely heavily on Xavier's existing behaviours, including path planning, navigation, vision and speech. ROGUE does not reason, for example, about which path the robot takes to reach a goal.

In Table 3, for example, the robot cannot deliver a particular item unless it (a) has the item in question, and (b) is in the correct location.

If any of the four preconditions are false, it will create a plan to achieve each of the preconditions. It takes the top level goal, (has-item <user> <item>), and selects an operator that will achieve it. It continues building the plan recursively, adding operators for each precondition that is not true in the state, until all of the operators in the leaf nodes have no untrue preconditions, yielding a network of plan steps and goals such as the one shown in Fig. 5.

In Table 4, we show how an operator represents that the robot will not go to a pickup location unless it needs to pickup an item there. It does not matter where the robot's current location is; the variable <old-room> is only instantiated when the robot arrives at the goal location.[1]

Other operators in the domain represent different task actions. By abstracting each request to the robot, such as which path the robot takes, ROGUE can more fully address issues arising from multiple interacting

Table 3. Item delivery operator.

```
(operator DELIVER-ITEM <room> <user> <item>
   (preconds (and (needs-item <user> <item>)
                  (robot-has-item <user> <item>)
                  (deliver-loc <user> <room>)
                  (robot-in-room <room>)))
   (effects ((add (has-item <user> <item>))
             (del (needs-item <user> <item>))
             (del (robot-has-item <user> <item>)))))
```

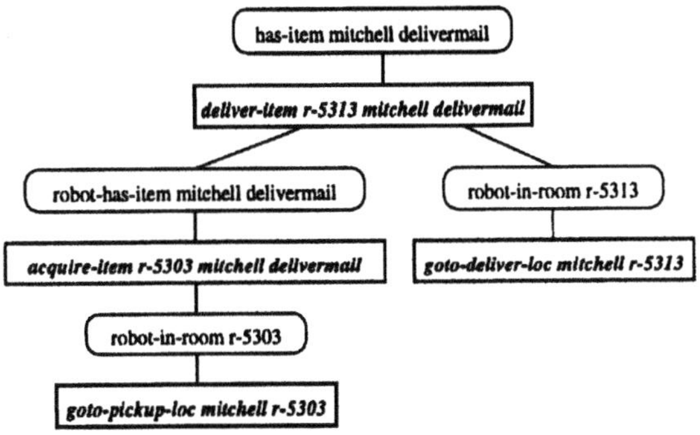

Figure 5. Plan for a single task problem. Goal nodes are shown in ovals, selected operators are shown in rectangles.

tasks, such as efficiency, resource contention, and reliability.

2.2.2. Search Control Rules. PRODIGY4.0 provides a method for creating *search control rules* that reduce the number of choices at each decision point in Table 2 by pruning the search space or suggesting a course of action while expanding the plan. In particular, control rules can select, prefer or reject a particular goal or action in a particular situation. Control rules can be used to focus planning on particular goals and towards desirable plans.

Table 4. Goto pickup location operator.

```
(operator GOTO-PICKUP-LOC <user> <room>
   (preconds
       (and (needs-item <who> <item>)
            (not (robot-has-item <who> <item>))
            (pickup-loc <who> <room>)))
   (effects ((<old-room> ROOM))
            ((del (robot-in-room <old-room>))
             (add (robot-in-room <room>)))))
```

Each time PRODIGY4.0 examines the set of unsolved pending goals, it fires its search control rules to decide which goal to expand. ROGUE interacts with PRODIGY4.0 by providing the set of control rules used to constrain PRODIGY4.0's decisions. Table 5 shows ROGUE's goal selection control rule that calls two functions, forcing PRODIGY4.0 to select the goals with high priority as well as the goals that can be opportunistically achieved (without compromising the main high-priority goal).

The function `(ancestor-is-top-priority-goal)` calculates whether the goal is required to solve a high-priority goal. ROGUE prioritizes goals according to a modifiable metric. For example, in the current implementation, this metric involves looking at the user's position in the department, the type of request and the deadline: *Priority* = *PersonRank* + *TaskRank* + *DeadlineRank*, where *DeadlineRank* is defined as shown in Fig. 6. (When the deadline is reached, the goal is removed from PRODIGY4.0's pending goals list; otherwise even a task of priority 0 would eventually be attempted after all other pending tasks have been completed.) This function could easily be replaced with alternatives (e.g., Williamson and Hanks, 1994).

The function `(compatible-with-top-priority-goal)` allows ROGUE to determine when different goals have similar features so that it can opportunistically achieve lower priority goals while achieving higher priority ones. For example, if multiple people whose offices are all in the same hallway asked for their mail to be picked up and brought to them, ROGUE would do all the requests in the same episode, rather

Table 5. Goal selection search control rule.

```
(control-rule SELECT-TOP-PRIORITY-AND-COMPATIBLE-GOALS
   (if (and (candidate-goal <goal>)
            (or (ancestor-is-top-priority-goal <goal>)
                (compatible-with-top-priority-goal <goal>))))
   (then select goal <goal>))
```

$$DeadlineRank = \begin{cases} \frac{R_{max}}{t_1 - t_0} \times (t - t_0) & t_0 \le t \le t_1 \\ 0 & \text{otherwise} \end{cases}$$

Where: t is the current time
R_{max} is the maximum possible rank value
t_1 = deadline − expected_execution_time
t_0 = deadline − 2× expected_execution_time

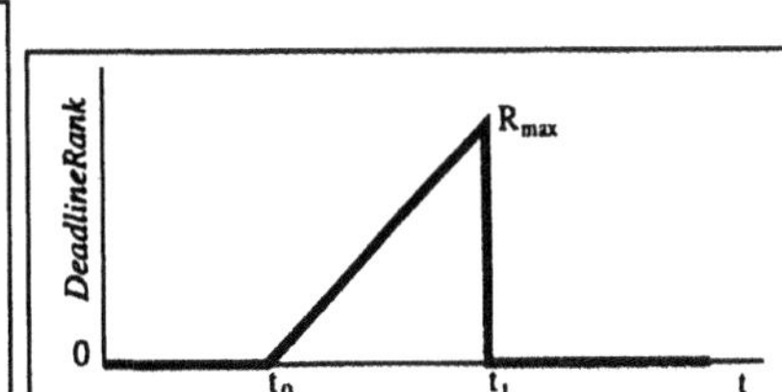

Figure 6. Calculating the priority rank of the deadline.

than only bringing the mail for the most important person. Compatibility is defined by physical proximity ("*on the path of*") with a fixed threshold for being too far out of the way.

It is possible that these rules will select too many compatible tasks, become "side-tracked," and therefore fail on the high-priority task. A preset threshold would serve as a pragmatic solution to this problem. We also do not deal with the issue of thrashing, i.e., receiving successively more important tasks resulting in no forward progress, because it has not been an issue in practice.

PRODIGY4.0 also uses a search control rule to select a good execution order of the applicable actions. It makes the choice based on an execution-driven heuristic which minimizes the expected total traveled distance from the current location.

2.3. *Suspending and Interrupting Tasks*

ROGUE needs to be able to respond quickly when new tasks arrive and also when priorities of existing tasks change. PRODIGY4.0 supports these changing objectives by making it easy to suspend and reactivate tasks.

PRODIGY4.0 grows the plan incrementally, meaning that each time it selects a goal to expand, the remainder of the plan is unaffected. The system can therefore easily suspend planning for one task while it plans for another. The planning already done for the suspended goals remains valid until PRODIGY4.0 is able to return to them.

When PRODIGY4.0 does in fact return to the suspended actions, it validates their preconditions in the state, expanding the plan if necessary, or continuing execution if appropriate.

Generally, the plans for the interrupted goals will not be affected by the planning and execution for the new goal. Occasionally, however, actions executed to achieve the new goal might undo or achieve parts of the interrupted plan. For example, the robot might have finished its new task in the pickup location of the interrupted task. There are also occasions in which exogenous events may change the state, such as if a user passed the robot in the corridor and took his mail at that time.

In cases like these, the execution monitoring algorithm will update PRODIGY4.0's state information and PRODIGY4.0 will know which preconditions it needs to re-achieve or to ignore. In Section 3.2 we discuss in more detail how side-effects of actions and exogenous events may affect interrupted or pending plans.

2.4. *Example*

We now present a detailed example of how PRODIGY4.0, ROGUE and Xavier interact in a two goal problem: `(has-item mitchell delivermail)` and `(has-item jhm deliverfax)`. The second (higher priority) goal arrives while ROGUE is executing the first action for the first goal.

Figure 7 shows the planning graph generated by PRODIGY4.0. We describe below the details of how it is generated. This example shows:

- how PRODIGY4.0 expands tasks,

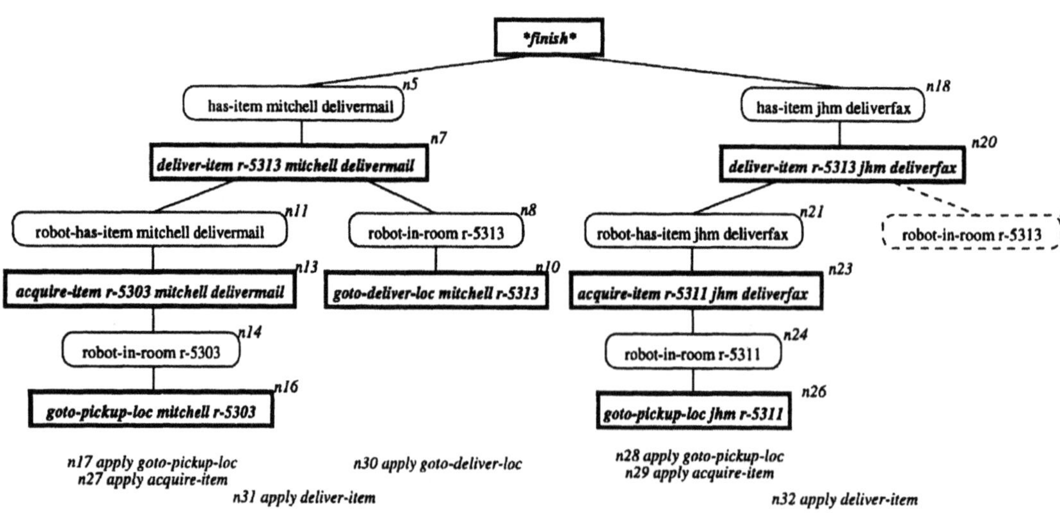

Figure 7. Plan for a two-task problem; goal nodes are in ovals, required actions are in rectangles.

- how search control rules affect PRODIGY4.0's selections, and
- how an asynchronous task request affects the plan.

We assume for the purposes of this example that no failures occur during execution. The example is perhaps overly detailed for a reader familiar with back-chaining planners; those readers could skip to the next section without loss of continuity.

We show the algorithmic sequence of steps of PRODIGY4.0. At each step, we show the lists of pending goals, *PG*, applicable operators, *Applicable-Ops*, and executed operators, *Executed-Ops*[2].

---

1. Request (has-item mitchell delivermail) arrives. ROGUE adds this goal to PRODIGY4.0's pending goals list, *PG*, and adds the following knowledge to PRODIGY4.0's state information:

```
(needs-item mitchell delivermail)
(pickup-loc mitchell r-5303)
(deliver-loc mitchell r-5313)
```

   PG is (has-item mitchell delivermail)
   Applicable-Ops is nil
   Executed-Ops is nil

2. PRODIGY4.0 fires its goal-selection search control rules, which selects this goal (node 5 of Fig. 7) as the highest priority goal (since it is the only choice). PRODIGY4.0 examines this goal to find an appropriate operator. It finds (DELIVER-ITEM <user> <room> <object>), and instantiates the variables: <user> := mitchell, <room> := r-5313 and <object> := delivermail, yielding the instantiated operator shown in node 7 of Fig. 7. Using means-ends analysis, PRODIGY4.0 identifies two preconditions not satisfied in the state: (robot-has-item mitchell delivermail) and (robot-in-room r-5313). PRODIGY4.0 adds these preconditions to the pending goals list.

   PG is (and (robot-has-item mitchell delivermail)
   (robot-in-room r-5313))
   Applicable-Ops is nil
   Executed-Ops is nil

3. PRODIGY4.0 continues expanding the plan for this task, yielding nodes 5 through 16. At this moment, two operators in the plan have all their preconditions met in the current state.

   PG is nil
   Applicable-Ops is (and (GOTO-DELIVER-LOC mitchell r-5313)
   (GOTO-PICKUP-LOC mitchell r-5303))
   Executed-Ops is nil

4. PRODIGY4.0 examines the set of *Applicable-Ops*, and based on ordering constraints (goal clobbering), selects (GOTO-PICKUP-LOC mitchell r-5303) to apply. ROGUE takes the applied operator, (GOTO-PICKUP-LOC mitchell r-5303) (node 17), and sends it to the robot for execution. (It does not need to verify preconditions in the real world since none can be changed by exogenous events.)

   PG is nil
   Applicable-Ops is (GOTO-DELIVER-LOC mitchell r-5313)
   Executed-Ops is nil

5. Request (has-item jhm deliverfax) arrives. ROGUE adds this goal to *PG*.
   ROGUE does not interfere with the currently executing action, namely (GOTO-PICKUP-LOC mitchell r-5303). Goodwin (1994) discusses methods to decide when to interfere.

```
PG is (has-item jhm deliverfax)
Applicable-Ops is (GOTO-DELIVER-LOC mitchell r-5313)
Executed-Ops is nil
```

6. The navigation module finally indicates completion of the action. ROGUE verifies the outcome (post-conditions) of the action, i.e., that it has arrived at the location `r-5313` (see Section 3 for a description). Now the action `(ACQUIRE-ITEM r-5303 mitchell delivermail)` is applicable.

```
PG is (has-item jhm deliverfax)
Applicable-Ops is (and (GOTO-DELIVER-LOC mitchell r-5313)
                       (ACQUIRE-ITEM r-5303 mitchell delivermail))
Executed-Ops is (GOTO-PICKUP-LOC mitchell r-5303)
```

7. PRODIGY4.0 fires ROGUE's search control rules, which select the new goal (since it is higher priority than the current task) (node 18). It expands the plan as above, except that instead of selecting a new operator to achieve the goal `(robot-in-room r-5313)`, it notices that the operator `(GOTO-DELIVER-LOC mitchell r-5313)` has the same effect, and does not redundantly add a new operator.

```
PG is nil
Applicable-Ops is (and (GOTO-DELIVER-LOC mitchell r-5313)
                       (ACQUIRE-ITEM r-5303 mitchell delivermail)
                       (GOTO-PICKUP-LOC jhm r-5311))
Executed-Ops is (GOTO-PICKUP-LOC mitchell r-5303)
```

8. PRODIGY4.0 selects `(ACQUIRE-ITEM r-5303 mitchell delivermail)` to apply (node 27); ROGUE verifies its preconditions and then sends it to the robot for execution. When the action is complete, ROGUE verifies the postconditions (i.e., that it now has mitchell's mail).

```
PG is nil
Applicable-Ops is (and (GOTO-DELIVER-LOC mitchell r-5313)
                       (GOTO-PICKUP-LOC jhm r-5311))
Executed-Ops is (GOTO-PICKUP-LOC mitchell r-5303)
                (ACQUIRE-ITEM r-5303 mitchell deliver-mail)
```

9. The execution constraint control rule now selects `(GOTO-PICKUP-LOC jhm r-5311)` as the next applicable operator (node 28). ROGUE sends it to Xavier for execution and monitors its outcome.

```
PG is nil
Applicable-Ops is (and (GOTO-DELIVER-LOC mitchell r-5313)
                       (ACQUIRE-ITEM r-5311 jhm deliverfax))
Executed-Ops is (GOTO-PICKUP-LOC mitchell r-5303)
                (ACQUIRE-ITEM r-5303 mitchell deliver-mail)
                (GOTO-PICKUP-LOC jhm r-5311)
```

10. ROGUE then acquires the fax.
11. ROGUE then goes to room 5313.
12. ROGUE delivers both items.

```
Solution:
    <GOTO-PICKUP-LOC mitchell r-5303>
  [arrival of second request]
    <ACQUIRE-ITEM r-5303 mitchell delivermail>
    <GOTO-PICKUP-LOC jhm r-5311>
    <ACQUIRE-ITEM r-5311 jhm deliverfax>
    <GOTO-DELIVER-LOC mitchell r-5313>
    <DELIVER-ITEM r-5313 jhm deliverfax>
    <DELIVER-ITEM r-5313 mitchell delivermail>
```

Figure 8. Final execution sequence.

The final execution order described in this example is shown in Fig. 8. This example illustrates the asynchronous handling of goals in ROGUE.

3. Execution and Monitoring

In this section we describe how ROGUE mediates the interaction between the planner and the robot. We show how symbolic action descriptions are turned into robot commands, as well as how robot sensor data is incorporated into the planner's knowledge base so that the planner can compensate for changes in the environment or unexpected failures of its actions.

The key to this communication model is based on a pre-defined language and model translation between PRODIGY4.0 and Xavier. The procedures to do this translation are manually generated, but are in a systematic format and may be extended at any time to augment the actions or sensing capabilities of the system. It is an open problem to automate the generation of these procedures because it is not only challenging to select what features of the world may be relevant for replanning, but also how to detect those features using existing sensors.

3.1. *Executing Actions*

Each action that PRODIGY4.0 selects must be translated into a form that Xavier will understand. ROGUE translates the high-level abstract action into a command sequence appropriate for execution.

PRODIGY4.0 allows arbitrary procedural attachments that are called during the operator application phase of the planning cycle (Stone and Veloso, 1996). Typically, we use these functions to give the planner additional information about the state of the world that might not be accurately predictable from the model of the environment. For example, this new information might show resource consumption or action outcomes.

ROGUE extends this information-gathering capability because, instead of *simulating* operator effects, ROGUE actually sends the commands to the robot for *real world* execution. Actually executing the planner's actions in this way increases system reliability and efficiency because the system can respond quickly to unexpected events such as failures and side effects, and combined planning and execution effort is reduced since actions are interleaved, and the planner knows the exact outcome of uncertain events.

In general, these procedures:

1. verify the preconditions of the operator,
2. execute the associated actions, and
3. verify the postconditions of the operator.

Some of these procedures also contain simple failure recovery procedures, particularly for actions that have common and known failures. For example, an action might simply be repeated, as in the `navigateToGoal` command. These procedures resemble schemas (Georgeff and Ingrand, 1989; Hormann et al., 1991) or RAPs (Firby, 1989; Gat, 1992; Pell et al., 1997), in that they specify how to execute the action, what to monitor in the environment, and some recovery procedures. ROGUE's procedures, however, do not contain complex recovery or monitoring procedures, such as when they have different costs or probabilities, since we feel that it is more appropriate for the planner to reason about when they should be used.

These command sequences may be executed directly by ROGUE (e.g., a command like `finger` to determine an office location), or sent via the TCA interface to the Xavier module designed to handle the command. The action `(ACQUIRE-ITEM <room> <user> <item>)`, for example, is mapped to a sequence of commands that allows the robot to interact with a human. The action `(GOTO-PICKUP-LOC <user> <room>)` is mapped to the commands shown in Table 6, extracted from an actual trace: (1) Announce intended action, (2) Ask Xavier's path planner to find the coordinates of a door near the room, (3) Navigate to those coordinates, and (4) Verify the outcome.

In the following section, we explain in more detail how ROGUE monitors the outcome of the action, and how failures may cause replanning or affect plans of interrupted tasks.

Table 6. The set of actions taken for executing the PRODIGY4.0 operator <GOTO-DELIVER-LOC mitchell r-5309>.

```
<GOTO-DELIVER-LOC MITCHELL R-5309>

SENDING COMMAND (tcaExecuteCommand "C_say" "Going to room 5309")
ANNOUNCING: Going to room 5309
SENDING COMMAND (tcaQuery "nearRoomQ" "5309")
   ...Query returned #(TASK-CONTROL::NEARROOMREPLY 567.0d0 3483.0d0 90.0d0)
SENDING COMMAND (tcaExpandGoal "navigateToG" #(TASK-CONTROL::MAPLOCDATA 567.0d0 3483.0d0))
   ...waiting...
   ...Action NAVIGATE-TO-GOAL finished (SUCCESS).
SENDING COMMAND (tcaQuery "visionWhereAmI")
   ...Query returned #(TASK-CONTROL::VISIONWHEREAMI "5309")
```

3.2. *Monitoring*

There are two types of events that ROGUE needs to monitor in the environment.

The first centers around *actions*. Each time ROGUE executes an action, it needs to verify its outcome because actions may have multiple outcomes or fail unexpectedly. ROGUE may need to invoke replanning, or select actions at a branching condition. ROGUE also needs to verify the preconditions of an action before executing it because the world may change, invalidating one of the system's beliefs. ROGUE uses a layered verification process, incrementally calling methods with greater cost and accuracy, until a predefined confidence threshold is reached. Action monitors are invoked only when the action is executed.

The second centers around exogenous events in the *environment*. Certain events may cause changes in the environment that affect current goals, or opportunities may arise that ROGUE can take advantage of. For example, ROGUE can monitor battery power, or examine camera images for open doors or particular objects. Environment monitors are invoked when relevant goals are introduced to the system.

Both types of monitoring procedures specify (1) what to monitor and (2) the methods that can be used to monitor it. *Action monitors* monitor the preconditions and effects of the action, while *environment monitors* are determined by the programmer. The action monitors, based on the planning domain model, provide a focus for execution monitoring. It is an open problem to autonomously decide what exogenous events to monitor that will be relevant for planning.

Although action monitoring is sequential and of limited time-span, while environment monitoring is parallel and continuous, the two sets of procedures have similar effects on planning.

Once ROGUE has done the required monitoring, ROGUE needs to update PRODIGY4.0's state description as appropriate. In execution monitoring, the update occurs when the object is detected, or when battery power falls below a certain threshold. In action monitoring, the critical update is when the *actual* outcome of the monitoring does not meet the *expected* outcome. These updates will force PRODIGY4.0 to re-examine its plan, adding or discarding operators as necessary.

If the primary effect of the action has been unexpectedly satisfied, ROGUE adds the knowledge to PRODIGY4.0's state description and PRODIGY4.0 does not attempt to achieve it. Observing the environment and maintaining a state description in this way improves the efficiency of the system because it will not attempt redundant actions.

If a required precondition is no longer true as a side-effect of some other action or environment monitoring, ROGUE deletes the relevant precondition from PRODIGY4.0's state. PRODIGY4.0 will therefore replan in an attempt to find an action that will re-achieve it.

In action monitoring, if the action fails, ROGUE will first try the built-in recovery methods. These recovery methods are very simple; more complex ones are treated as separate operators for PRODIGY4.0 to reason about. For example, ROGUE will try calling the navigation routine a predefined number of times before deciding that the action completely failed. At the scene of a pickup or delivery, if ROGUE times-out while waiting for a response to a query, ROGUE will prompt for a user a second time before failing. If, despite the built-in recovery methods, ROGUE determines that the action has completely failed, ROGUE will delete the effect from PRODIGY4.0's state description, and PRODIGY4.0 will replan to achieve it.

Occasionally during environment monitoring, knowledge will unexpectedly be added to the state that causes an action to become executable, or a task to become higher priority. Each time PRODIGY4.0 makes a decision, it re-examines all of its options, and will factor the new action or goal into the process.

In this manner, ROGUE is able to detect execution failures and compensate for them, as well as to respond to changes in the environment. The interleaving of planning and execution reduces the need for replanning during the execution phase and increases the likelihood of overall plan success because the planner is constantly being updated with information about changes in the world. It allows the system to adapt to a changing environment where failures can occur.

3.3. *Example of how ROGUE Handles Failures*

One of Xavier's actions that ROGUE monitors is the `navigateToG(oal)` command, used by both the `(GOTO-PICKUP-LOC <user> <room>)` and the `(GOTO-DELIVER-LOC <user> <room>)` operators. `navigateToG` reports a success when the robot arrives at the requested goal. `navigateToG` may fail under several conditions, including detecting a bump, corridor or door blockage, or lack of forward progress. The module is able to autonomously compensate for certain problems, such as obstacles and missing landmarks. Navigation is done using Partially Observable Markov Decision Process models (Simmons and Koenig, 1995), and the inherent uncertainty of this probabilistic model means that the module may occasionally report success even when it has not actually arrived at the desired goal location.

When `navigateToG` reports a failure or a low-probability success, ROGUE verifies the location. ROGUE first tries to verify the location autonomously, using its cameras. The vision module looks for a door in the general area of the expected door, and finds the room label, and reads it. If this module fails to find a door, fails to find a label, or returns low confidence in its template matching, ROGUE falls back to a second verification procedure, namely using the speech module to ask a human. We assume that verification step gives complete and correct information about the robot's actual location; other researchers are focussing on the open problem of sensor reliability (Hughes and Ranganathan, 1994; Thrun, 1996).

If ROGUE detects that in fact the robot is *not* at the correct goal location, ROGUE updates the navigation module with the new information and re-attempts to navigate to the desired location. If the robot is still not at the correct location after a constant number of attempts (three in our current implementation), ROGUE updates PRODIGY4.0's task knowledge to reflect the robot's actual position, rather than the expected position.

In general, PRODIGY4.0 has several different operators that can achieve a particular effect, and will successfully replan for the failure. In this case, however, there are no other alternative methods of navigating, and PRODIGY4.0 declares that the task cannot be successfully achieved.

3.4. *Example of how ROGUE Handles Side-effects*

Occasionally, suspending one task for a second one will mean that work done for the first will be undone by work done for the second. ROGUE needs to detect these situations and plan to re-achieve the undone work. Consider a simple situation that illustrates this re-planning process:

Task one:	**Task two:**
1a. goto 5301	2a. goto 5409
1b. pick up mail	2b. pick up fed-ex package
1c. goto 5315	2c. goto 4320
1d. drop off mail	2d. drop package off

Many possible interleaved planning and execution scenarios may occur; below are two possibilities:

- **[Normal:]** Executes 1a and 1b. While executing, the request for two arrives. ROGUE decides that task two is more important. Task one is suspended; step 1c is pending. Plans for and executes task two. Returns to step 1c, verifies that it is still needed to complete the task and can still be done, then does 1c and 1d.
- **[Undone Action:]** Executes 1a. While executing 1b, the request for task two arrives. 1b times-out, indicating that the mail-room person was not there to give the robot the mail. ROGUE decides task two is more important, and suspends task one; step 1b is pending. ROGUE plans for and executes task two. PRODIGY4.0 returns to step 1b, discovers that a precondition is not true: `(robot-in-room <5301>)`. PRODIGY re-plans to achieve it, and then re-executes step 1a, and then finishes the task as expected.

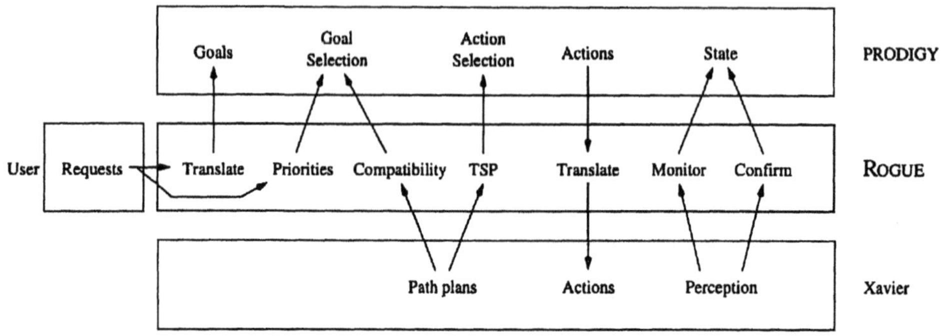

Figure 9. What ROGUE does.

4. Conclusion

ROGUE is fully implemented and operational. The system completes all requested tasks, running errands between offices in our building. Execution results are presented in detail elsewhere (Simmons et al., 1997).

We have presented one aspect of ROGUE, an integrated planning and execution robot architecture. We have described how PRODIGY4.0 gives ROGUE the power

- to integrate asynchronous requests,
- to prioritize goals,
- to suspend and reactivate tasks,
- to recognize compatible tasks and opportunistically achieve them,
- to execute actions in the real world, integrating new knowledge which may help planning, and
- to monitor and recover from failure.

ROGUE represents a successful integration of a classical artificial intelligence planner with a real mobile robot. The complete planning and execution cycle for a *single* task can be summarized as follows:

1. ROGUE receives a task request from a user.
2. ROGUE requests a plan from PRODIGY4.0.
3. PRODIGY4.0 generates a plan and passes executable steps to ROGUE.
4. ROGUE translates and sends the planning steps to Xavier.
5. ROGUE monitors execution and identifies goal status; in case of failure, PRODIGY4.0's state information domain modified and PRODIGY4.0 will replan for decisions.

ROGUE handles multiple goals, interleaving the individual plans to maximize overall execution efficiency.

Figure 9 summarizes the information exchanged between the user, PRODIGY4.0, and Xavier under ROGUE's mediation. ROGUE constrains PRODIGY4.0's decisions through calculations on task priority, task compatibility, and execution efficiency. ROGUE translates PRODIGY4.0's symbolic action descriptions into Xavier commands, and also translates Xavier's perception information into PRODIGY4.0 domain description.

The contributions of our work to the Xavier project are in the high-level reasoning parts of the system, allowing the robot to efficiently handle multiple, asynchronous interacting goals, and to effectively interleave planning and execution in a real world system. Execution monitoring based on a planning model allows the systematic identification of environment monitors.

Interleaving planning with execution enhances a deliberative robot system in numerous ways. One such benefit is that the system can sense the world to acquire necessary domain knowledge in order to continue planning. For example, it can ask directions, look to see if doors are open or closed, or check whether it needs to recharge its batteries. Another benefit is reduced planning effort because the system does not need to plan for all possible failure contingencies; instead, it can execute an action to find out its actual outcome.

ROGUE advances the state of the art of the integration of planning and execution in robotic agent. In a unique novel way, ROGUE is designed as the integration of two independently developed platforms. PRODIGY4.0 is a general-purpose planner and Xavier can be viewed as a general-purpose navigational robot. ROGUE merges the functionality of these two systems in a real implementation that demonstrates the feasibility of connecting both systems in a rich task environment, namely the achievement of asynchronous user requests. (ROGUE therefore also shows how the PRODIGY4.0 planner and the TCA approach in Xavier are in fact robust architectures.)

Strictly looking at ROGUE only from the viewpoint of the integration of planning and execution, ROGUE compares well with other special-purpose systems such as NMRA and 3_T. Given the general-purpose character of the PRODIGY4.0 planner, ROGUE could easily be applied to other executing platforms and tasks by a flexible change of PRODIGY4.0's specification of the domain.

The goal of our research is to build a complete planning, executing and learning autonomous robotic agent. ROGUE's contributions go beyond the integration of planning and execution. ROGUE incorporates learning from execution experience (Haigh and Veloso, 1998). The learning algorithm involves extracting relevant information from real execution traces in order to detect patterns in the environment to improve the robot's behaviour. The ability to learn task-relevant knowledge conveniently matches PRODIGY4.0's search control representation, and learned situational-dependent arc costs can be incorporated into Xavier's route planner knowledge. Through its interleaved planning and execution behaviour, ROGUE provides an appropriate platform to collect the required execution data for learning.

Acknowledgments

The authors would like to thank Eugene Fink, Sven Koenig, Illah Nourbakhsh, Joseph O'Sullivan, Gary Pelton and the anonymous reviewers for feedback on this article. We would also like to thank the members of the Xavier and PRODIGY groups for feedback, comments and criticism on our research.

This research is sponsored in part by (1) the National Science Foundation under Grant No. IRI-9502548, (2) by the Defense Advanced Research Projects Agency (DARPA), and Rome Laboratory, Air Force Materiel Command, USAF, under agreement number F30602-95-1-0018, (3) the Natural Sciences and Engineering Council of Canada (NSERC), and (4) the Canadian Space Agency (CSA). The views and conclusions contained herein are those of the authors and should not be interpreted as necessarily representing the official policies or endorsements, either expressed or implied, of the NSF, DARPA, Rome Laboratory, the U.S. Government, NSERC or the CSA.

Notes

1. The representation of the operators, for example (GOTO-PICKUP-LOC) and (GOTO-DELIVER-LOC), is not intrinsic to the task, but it can be relevant to planning efficiency. We have an implementation of the domain with a single (GOTO-LOC) operator with less constrained preconditions, which leads to more backtracking while the planner selects the correct order of desired locations. We can also create a search control rule to guide the planning choices; this is logically equivalent to separating the operators, but with some additional match cost.
2. For readers more familiar with PRODIGY literature, the *Executed-Ops* correspond to PRODIGY4.0's *head-plan*, while the plan shown in Fig. 7 corresponds to PRODIGY4.0's *tail-plan*.

References

Ambros-Ingerson, J.A. and Steel, S. 1988. Integrating planning, execution and monitoring. In *Proceedings of the Seventh National Conference on Artificial Intelligence, AAAI-88*, St. Paul, MN, AAAI Press: Menlo Park, CA, pp. 83–88.

Atkins, E.M., Durfee, E.H., and Shin, K.G. 1996. Detecting and reacting to unplanned-for world states. In *Papers from the 1996 AAAI Fall Symposium "Plan Execution: Problems and Issues,"* Boston, MA, AAAI Press: Menlo Park, CA, pp. 1–7.

Blythe, J. 1994. Planning with external events. In *Proceedings of the Tenth Conference on Uncertainty in Artificial Intelligence*, Seattle, WA, Morgan Kaufmann: San Mateo, CA, pp. 94–101.

Bonasso, R.P. and Kortenkamp, D. 1996. Using a layered control architecture to alleviate planning with incomplete information. In *Proceedings of the AAAI Spring Symposium "Planning with Incomplete Information for Robot Problems,"* Stanford, CA, AAAI Press: Menlo Park, CA, pp. 1–4.

Carbonell, J.G., Knoblock, C.A., and Minton, S. 1990. PRODIGY: An integrated architecture for planning and learning. In *Architectures for Intelligence*, K. VanLehn (Ed.), Erlbaum: Hillsdale, NJ. Also available as Technical Report CMU-CS-89-189, Computer Science Department, Carnegie Mellon University, Pittsburgh, PA.

Dean, T.L. and Boddy, M. 1988. An analysis of time-dependent planning. In *Proceedings of the Seventh National Conference on Artificial Intelligence, AAAI-88*, St. Paul, MN, AAAI Press: Menlo Park, CA, pp. 49–54.

Dean, T., Basye, K., Chekaluk, R., Hyun, S., Lejter, M., and Randazza, M. 1990. Coping with uncertainty in a control system for navigation and exploration. In *Proceedings of the Eigth National Conference on Artificial Intelligence, AAAI-90*, Boston, MA, MIT Press: Cambridge, MA, pp. 1010–1015.

DellaFera, C.A., Eichin, M.W., French, R.S., Jedlinsky, D.C., Kohl, J.T., and Sommerfeld, W.E. 1988. The Zephyr notification service. In *Proceedings of the USENIX Winter Conference*, Dallas, TX, USENIX Association: Berkeley, CA, pp. 213–219.

Drummond, M., Swanson, K., Bresina, J., and Levinson, R. 1993. Reaction-first search. In *Proceedings of the Thirteenth International Joint Conference on Artificial Intelligence, IJCAI-93*, Chambery, France, Morgan Kaufmann: San Mateo, CA, pp. 1408–1414.

Fikes, R.E., Hart, P.E., and Nilsson, N.J. 1972. Learning and executing generalized robot plans. *Artificial Intelligence*, 3(4):231–249.

Firby, R.J. 1989. Adaptive execution in complex dynamic worlds. Ph.D. Dissertation, Yale University, New Haven, CT.

Gat, E. 1992. Integrating planning and reacting in a heterogeneous asynchronous architecture for controlling real-world mobile robots. In *Proceedings of the Tenth National Conference on Artificial Intelligence, AAAI-92*, San Jose, CA, AAAI Press: Menlo Park, CA, pp. 809–815.

Georgeff, M.P. and Ingrand, F.F. 1989. Decision-making in an embedded reasoning system. In *Proceedings of the Eleventh International Joint Conference on Artificial Intelligence, IJCAI-89*, Detroit, MI, Morgan Kaufmann: San Mateo, CA, pp. 972–978.

Gervasio, M.T. and DeJong, G.F. 1991. Learning probably completable plans. Technical Report UIUCDCS-R-91-1686, University of Illinois at Urbana-Champaign, IL, Urbana, IL.

Goodwin, R. 1994. Reasoning about when to start acting. In *Artificial Intelligence Planning Systems: Proceedings of the Second International Conference, AIPS-94*, Chicago, IL, K. Hammond (Ed.), AAAI Press: Menlo Park, CA, pp. 86–91.

Goodwin, R. and Simmons, R.G. 1992. Rational handling of multiple goals for mobile robots. In *Artificial Intelligence Planning Systems: Proceedings of the First International Conference, AIPS-92*, College Park, MD, J. Hendler (Ed.), Morgan Kaufmann: San Mateo, CA, pp. 86–91.

Haigh, K.Z. and Veloso, M. 1996. Interleaving planning and robot execution for asynchronous user requests. In *Proceedings of the International Conference on Intelligent Robots and Systems, IROS*, Osaka, Japan, IEEE Press: New York, NY, pp. 148–155.

Haigh, K.Z. and Veloso, M.M. 1997. High-level planning and low-level execution: Towards a complete robotic agent. In *Proceedings of the First International Conference on Autonomous Agents*, Marina del Rey, CA, W.L. Johnson (Ed.), ACM Press: New York, NY, pp. 363–370.

Haigh, K.Z. and Veloso, M.M. 1998. Learning situation-dependent costs: Improving planning from probabilistic robot execution. In *Proceedings of the Second International Conference on Autonomous Agents*, Minneapolis, MN, K.P. Sycara (Ed.), AAAI Press: Menlo Park, CA. To appear.

Hormann, A., Meier, W., and Schloen, J. 1991. A control architecture for and advanced fault-tolerant robot system. *Robotics and Autonomous Systems*, 7(2–3):211–225.

Hughes, K. and Ranganathan, N. 1994. Modeling sensor confidence for sensor integration tasks. *International Journal of Pattern Recognition and Artificial Intelligence*, 8(6):1301–1318.

Kushmerick, N., Hanks, S., and Weld, D. 1993. An algorithm for probabilistic planning. Technical Report 93-06-03, Department of Computer Science and Engineering, University of Washington, Seattle, WA.

Lyons, D.M. and Hendriks, A.J. 1992. A practical approach to integrating reaction and deliberation. In *Artificial Intelligence Planning Systems: Proceedings of the First International Conference, AIPS-92*, College Park, MD, J. Hendler (Ed.), Morgan Kaufmann: San Mateo, CA, pp. 153–162.

Mansell, T.M. 1993. A method for planning given uncertain and incomplete information. In *Proceedings of the Ninth Conference on Uncertainty in Artificial Intelligence*, Washington, DC, Morgan Kaufmann: San Mateo, CA, pp. 250–358.

McDermott, D. 1992. Transformational planning of reactive behavior. Technical Report YALE/CSD/RR#941, Computer Science Department, Yale University, New Haven, CT.

Nilsson, N.J. 1984. Shakey the robot. Technical Report 323, AI Center, SRI International, Menlo Park, CA.

Nourbakhsh, I. 1997. Interleaving planning and execution for autonomous robots, Dordrecht, Netherlands: Kluwer Academic. Ph.D. thesis. Also available as technical report STAN-CS-TR-97-1593, Department of Computer Science, Stanford University, Stanford, CA.

O'Sullivan, J., Haigh, K.Z., and Armstrong, G.D. 1997. Xavier. Carnegie Mellon University, Pittsburgh, PA. Manual, Version 0.3, unpublished internal report. Available via `http://www.cs.cmu.edu/~Xavier/`.

Pell, B., Bernard, D.E., Chien, S.A., Gat, E., Muscettola, N., Nayak, P.P., Wagner, M.D., and Williams, B.C. 1997. An autonomous spacecraft agent prototype. In *Proceedings of the First International Conference on Autonomous Agents*, Marina del Rey, CA, W.L. Johnson (Ed.), ACM Press: New York, NY, pp. 253–261.

Pryor, L.M. 1994. Opportunities and planning in an unpredictable world, Ph.D. Dissertation, Northwestern University, Evanston, Illinois. Available as Technical Report number 53.

Schoppers, M.J. 1989. Representation and automatic synthesis of reaction plans, Ph.D. Dissertation, Department of Computer Science, University of Illinois, Urbana-Champaign, IL. Available as Technical Report UIUCDCS-R-89-1546.

Simmons, R. 1994. Structured control for autonomous robots. *IEEE Transactions on Robotics and Automation*, 10(1):34–43.

Simmons, R. and Koenig, S. 1995. Probabilistic robot navigation in partially observable environments. In *Proceedings of the Fourteenth International Joint Conference on Artificial Intelligence, IJCAI-95*, Montréal, Québec, Canada, Morgan Kaufmann: San Mateo, CA, pp. 1080–1087.

Simmons, R., Goodwin, R., Haigh, K.Z., Koenig, S., and O'Sullivan, J. 1997. A layered architecture for office delivery robots. In *Proceedings of the First International Conference on Autonomous Agents*, W.L. Johnson (Ed.), Marina del Rey, CA, ACM Press: New York, NY, pp. 245–252.

Stone, P. and Veloso, M.M. 1996. User-guided interleaving of planning and execution. In *New Directions in AI Planning*, Amsterdam, IOS Press: Netherlands, pp. 103–112.

Thrun, S. 1996. A Bayesian approach to landmark discovery and active perception for mobile robot navigation. Technical Report CMU-CS-96-122, School of Computer Science, Carnegie Mellon University, Pittsburgh, PA.

Veloso, M.M., Carbonell, J., Pérez, M.A., Borrajo, D., Fink, E., and Blythe, J. 1995. Integrating planning and learning: The PRODIGY architecture. *Journal of Experimental and Theoretical Artificial Intelligence*, 7(1):81–120.

Williamson, M. and Hanks, S. 1994. Optimal planning with a goal-directed utility model. In *Artificial Intelligence Planning Systems: Proceedings of the Second International Conference, AIPS-94*, Chicago, IL, K. Hammond (Ed.), AAAI Press: Menlo Park, CA, pp. 176–180.

Karen Zita Haigh is currently completing her Ph.D. in Computer Science at Carnegie Mellon University in Pittsburgh, Pennsylvania. Her undergraduate degree was completed in 1992 at the University of Ottawa in Ottawa, Ontario, Canada. Her thesis is a robot learning system that uses feedback from execution experience to improve efficiency of generated plans. It creates situation-dependent costs so that plans are tailored to particular situations, and is used on Xavier's route planner and in ROGUE. She also built analogical reasoning system to automatically generate high-quality routes in a city map. Her research interests include planning, machine learning, and robotics.

Manuela M. Veloso is a Finmeccanica Associate Professor in the Computer Science Department at Carnegie Mellon University. She received her Ph.D. in Computer Science from CMU in 1992. Dr. Veloso received the NSF Career Award and was the recipient of the Finmeccanica Chair in 1995. In 1997, she was awarded the Allen Newell Excellence in Research Award by the School of Computer Science at CMU. Dr. Veloso is the author of a monograph on "Planning by Analogical Reasoning." She co-edited two books, "Symbolic and Visual Learning" and "Topics of Case-based Reasoning". Dr. Veloso's research involves the integration of planning, execution and learning in dynamic environments, and in particular with multiple agents. She investigates memory-based machine learning techniques for the processing and reuse of problem experience.

Autonomous Robots 5, 97–110 (1998)

Integrated Premission Planning and Execution for Unmanned Ground Vehicles

EDMUND H. DURFEE, PATRICK G. KENNY AND KARL C. KLUGE
EECS Department, University of Michigan, Ann Arbor, MI 48109
durfee@umich.edu
pkenny@umich.edu
kckluge@umich.edu

Abstract. Fielding robots in complex applications can stress the human operators responsible for supervising them, particularly because the operators might understand the applications but not the details of the robots. Our answer to this problem has been to insert agent technology between the operator and the robotic platforms. In this paper, we motivate the challenges in defining, developing, and deploying the agent technology that provides the glue in the application of tasking unmanned ground vehicles in a military setting. We describe how a particular suite of architectural components serves equally well to support the interactions between the operator, planning agents, and robotic agents, and how our approach allows autonomy during planning and execution of a mission to be allocated among the human and artificial agents. Our implementation and demonstrations (in real robots and simulations) of our multi-agent system shows significant promise for the integration of unmanned vehicles in military applications.

Keywords: cooperating robots, coordination, multi-agent system

Introduction

Keeping people out of harm's way is a driving motivation for the development of unmanned ground vehicles (UGVs). UGVs (Fig. 1) can operate in hazardous environments, such as in contaminated areas or behind enemy lines on battlefields, conducting reconnaissance, surveillance, and target acquisition (RSTA) on behalf of distant human operators. Specifically, in this paper, we will consider the development of UGVs for scouting missions in military settings.

Fielding UGVs for scouting missions means that the UGVs will be placed under the control not of roboticists, but of soldiers who are commanding a variety of human and mechanized systems. This imposes significant constraints on the development of UGVs and the tools for their control. First, because an operator is a soldier, UGVs need to be tasked in military terms, similarly to how manned scouting vehicles would be tasked, rather than in robotic terms. Second, because an operator is simultaneously supervising various systems, the operator's attention cannot be devoted to teleoperating the UGVs; instead, the UGVs need to be able to operate with minimal intervention by the operator. Third, because lines of communication are uncertain, an operator is only sporadically available, and cannot be depended upon for prompt supervision and redirection.

These factors motivate the development of UGVs, and operator interfaces, that internalize substantial military knowledge, including standard procedures for achieving objectives, and that know the current objectives of an operator. With this knowledge, the overall unmanned system can be migrated from a "man-must-be-in-the-loop" to a "man-can-be-in-the-loop" strategy, such that robust and adaptive operation is not entirely dependent on the man-in-the-loop.

Toward this end, our research has treated UGVs as semi-autonomous agents that are capable of sensing their environment and adopting plans from their repertoire that are most suitable for achieving their objectives given the circumstances. Because the UGVs are often

*A preliminary version of this paper was presented at the First International Conference on Autonomous Agents in February 1997.

Figure 1. UGVs.

tasked in teams, our work has emphasized techniques for improving a UGV's awareness of the more global circumstances and for coordinating the plans of UGVs. Moreover, because they are also teamed with a human, the UGVs need to treat the human (or, more properly, the human-interface captured in the operator workstation (OWS)) as an agent as well, which also is adopting plans (mission specifications) from its repertoire in response to input from the UGVs and from the operator.

Rather than treat these loci of activity in a piecemeal fashion, the approach we outline here asserts that permeating all of these tasks, ranging from the specification of missions down to the invocation of robotic actions, is knowledge about how to pursue these tasks that can be combined in flexible, situation specific ways. Thus, in this paper, we describe how a procedural encoding of domain knowledge and mechanisms to use it provide a unifying framework for accomplishing the goals of (1) premission planning, (2) contingency preplanning, (3) reactive doctrinally correct response to unanticipated events, and (4) on-the-fly semi-autonomous coordination and replanning.

We have realized these goals by adapting procedural reasoning techniques (Georgeff and Lansky, 1986), embodied in the University of Michigan Procedural Reasoning System (UMPRS), to the planning and control of activities within, and across, UGVs and the operator (Lee et al., 1994). UMPRS controls a UGV based on its sensed situation and on the directives and objectives handed down from the operator's interface agent. UMPRS also controls the OWS by retrieving mission plans and taking actions to prompt the operator for refinements to objectives and mission parameters. Thus, the military intent and knowledge is distributed across the operator, the OWS, and the UGVs, and each of these can prompt another to allow mixed-initiative development of situation-specific mission definition, elaboration, planning, and revision.

In this paper, we outline the development constraints that we have faced in generating our agent processes for the UGV domain, and then describe our solution to the problems in terms of supporting premission plan definition and elaboration, plan execution across multiple agents, and runtime replanning during mission execution. Our overall solution comprises an integrated mission planning and execution system that is inherently distributed among multiple agents that embody explicit goals, plans, and beliefs. We also evaluate our solution both using a real robotic vehicle in a limited setting, and using a standard simulation system for demonstrating all of the components of our system working together for the planning and control of four UGVs by a single operator.

Problem Definition

The challenges in deploying UGVs in military settings fall into two broad categories: sociological and technological. From a sociological viewpoint, gaining acceptance for advanced technologies, and in particular technologies that strive for some level of autonomy, is problematic in a rigidly defined community such as the military. Throughout the UGV project, reactions by military personnel to the idea of UGVs that could improvise (within doctrine) so as to persistently achieve objectives have been in a bimodal distribution centered on "enthused" on the one hand and "horrified" on the other, with few moderate opinions between. The conservative default has been to strongly rely on the operator, and to provide vehicles with enough flexibility to dodge obstacles, but not to decide on new destinations, and certainly not to fire a weapon! Introducing agent technology, which inherently makes agents more flexible and persistent in achieving their goals, therefore has been an uphill battle.

From a technological viewpoint, several roadblocks have stood in the way of developing deployable UGVs. One is the limited sensing and mobility capabilities of real unmanned vehicles. Over the last several years, significant strides have been made in terms of mobility, such that a UGV can move among destinations off-road, at least in a relatively benign environment (without trenches, wire fences, etc.). Perception has lagged behind to an extent, not surprisingly. That is, mobility concentrates perception on aspects of the world that impact traversability; more generally, situation awareness requires perception that is much broader and richer. From the perspective of autonomous UGVs, rich perception is critical, since most (if not all) autonomy is manifested in terms of responding competently to unanticipated situations.

A second technological roadblock has been the constraints imposed by legacy software and hardware components. After substantial investment in robotic control systems and user interface tools, a project sponsor is anxious to build from these, even if it ties the hands of the current developers. Thus, to a significant extent, the development of agent technologies to be embedded in UGVs and in the OWS was circumscribed by having to weave together capabilities inherent in the legacy systems.

In summary, then, the problem of developing autonomous UGVs for military use can be viewed as generating decentralized agent technologies that could interface to military users on one end and robotic vehicles on the other such that the degree of autonomous control possessed by an agent (operator, OWS, UGV) can be flexibly changed as attitudes and acceptance changes. This is summarized in Fig. 2.

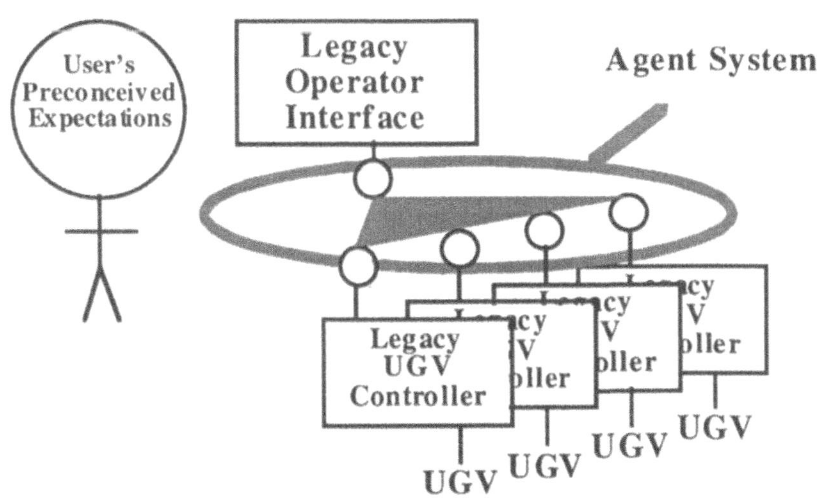

Figure 2. Agent system bridges gap.

Premission Planning

The first hurdle to overcome in bringing robotic technology into a military arena is to allow people who think in military terms to define a mission for the robotic systems without having to understand details of how the robots work. The premission planning capability supports the operator by assisting, and sometimes automating, the translation and elaboration of military objectives into robotic actions.

At the heart of premission planning is the UMPRS agent architecture, a C++ implementation of many of SRI's procedural reasoning concepts (Georgeff and Lansky, 1986), tailored specifically to support the development of agent technologies for operator assistance and autonomous execution. UMPRS supports specification of a mission's "operation order" in terms of a sequence of military procedures, each performed within the context of a larger set of objectives and constraints. Thus, whereas the robotic plan executed directly on the vehicles consists of a sequence of primitive commands, it is within the UMPRS system that the actual intent of the mission, along with the expected (and possibly alternative) strategies for its accomplishment, are represented. It is because of this more complete knowledge that the UMPRS-based premission planner is able to assist the operator: decomposing the mission into more detailed segments, possibly formulating some portions of actual robotic plans, making suggestions to the operator about features of the mission plan, and prompting the operator for further input to resolve ambiguities in the specification.

The premission planner works by accepting as input a map, generated by the (legacy) mission specification interface, that contains icons of the military measures of interest for the mission. The icons could have been entered in an arbitrary order, and the sequence and strategies for pursuing the measures might not be obvious strictly from the map. As a result, in order to resolve ambiguities the premission planner draws on its library of templates for operations, or can call on the operator to provide such a template. By matching the military symbology to the template, the premission planner begins formulating the sequence of military actions that must be carried out by the robotic vehicles. If matches are ambiguous, the interface can confirm its interpretation of the mission with the operator. If matches are impossible, the interface can prompt the operator to supply additional mission control measures, or can indicate to the operator why existing measures are inappropriate for the mission. It is important to note, moreover, that the procedures by which the mission planner's interface interacts with the operator are also explicitly represented and executed, meaning that the interface can be tailored to the needs and preferences of different operators.

As it elaborates the mission, the planner can call on more specialized tools to formulate portions of the robotic plan. These include tools for route planning (Stentz, 1995), for planning RSTA observation points (Cook et al., 1996; Kluge et al., 1994), and for planning formations (Balch and Arkin, 1995). The premission planner incorporates the information returned by these tools into its representation of the sequence of plan steps.

The final sequence of plan steps formulated by the premission planner can entirely dictate the control actions that the UGVs will invoke throughout the mission. These can be displayed to the operator (if the operator can understand them), and can be directly edited by the operator. So, in the case where the automated system is given no autonomy, the operator has the final say about precisely the robotic actions to be taken by the UGVs. If the operator relinquishes this control, the premission plan at the robotic level can be downloaded to the UGVs to be executed verbatim, meaning that autonomy is restricted to the operator and premission planner (on the OWS). However, robust performance needs dictate that the premission planner only elaborate the plan steps to an intermediate level of detail, and pass the resulting sequence of subgoals (with constraints on their accomplishment) to the UGVs, which will each then elaborate further to develop the detailed control plan. Our architecture makes no commitment to which of these different degrees of autonomy is instantiated at any given time; rather, it will depend on the particular knowledge and procedures provided to the various agents.

Contingency Preplanning

The intrinsic limitations in the accuracy with which actions can be carried out in the world, and the fact that the environment is likely to differ from what is expected, make it unlikely that a robotic vehicle can flawlessly execute a preplanned mission without complications. As a result, the UGV should be able to recover from deviations from its expectations in a variety of ways. Some deviations might involve straightforward recovery actions; for example, if a road is barricaded, the vehicle might go offroad around the barricade and then return to the road and its original plan. Other deviations might require some immediate response; for example, if fired upon, the UGV should move to cover reflexively. But other deviations might require more deliberation; for example, if it finds that a bridge it was to traverse is gone, a vehicle would need to consider the heightened likelihood of enemy contact near such an obstacle to navigation, alternative routes to reach its original objective, and the degree to which it can still achieve its mission goals by taking those routes.

Contingency preplanning deals with the more reflexive responses that a vehicle might take, while more involved changes in plans is in the province of replanning techniques. The idea behind contingency preplanning is straightforward: before the UGV begins executing its plan, the planner should anticipate what might go wrong from a tactical standpoint, and should preprogram responses for those deviations so that the vehicle can recover without hesitation. The challenges in contingency preplanning, therefore, are in developing mechanisms by which the mission planner can anticipate likely deviations, and can plan responses in a "what if" type of mode.

We have operationalized these mechanisms as part of the UMPRS mission planning capability by extending UMPRS's representations of standard operating procedures to model the expected/intended effects of such procedures. With this additional information, we can run UMPRS in a "forward simulation" mode to capture the expected sequences of actions that will occur so as to evaluate a possible ("what if") course of action before it is actually committed to. These same capabilities support the anticipation of contingencies, since the predicted results of a procedure can also capture the procedure's possible faults or side effects.

Moreover, we can associate particular contingencies with particular primitive procedures. For example, a common contingency in following a road might be that the road is obstructed; thus, whenever the planner inserts a step to follow a road, it automatically includes the contingency plan for dealing with an obstructed road. In fact, possible contingencies can be associated not only with primitive plan steps, but also for more extensive steps, such as the contingency of receiving fire that could be associated with a larger sequence of operations behind enemy lines.

Having identified a deviated state into which a vehicle might be thrown, the planner must generate a robotic plan for responding to that state. Thanks to the fact that the intent of the mission is captured in the UMPRS-based planner, the contingency preplanning capabilities can search for procedures to invoke that robustly pursue the mission intent, including actions that remove the deviation so that the original plan can be resumed. Because the intent also can capture high-level goals (such as survival), the contingency preplanner can also suggest responses that abort portions of a mission in favor of these other goals.

Contingency plans are elaborated by the same mechanisms as premission planning, including prompting the operator for information if needed (such as "If attacked, where should the vehicles fall back to?") and invoking more specialized tools (such as to determine a formation to use if surprised by an enemy). The contingency plans are woven into the original mission plan as conditional branches.

Vehicular Planning and Execution

Depending on the level of autonomy assigned to the vehicles, they might engage in further premission planning prior to executing the mission. As discussed above, at the premission phase, the operator works with the interface provided by the OWS to specify and elaborate the mission. It could be that the OWS serves only as an editing tool, such that the entire robotic mission is specified directly by the operator. Or the operator and OWS could cooperatively formulate the detailed robotic plan. In such cases, a UGV simply receives a sequence of commands in its (legacy) robotic control language, and blindly executes these until done or until interrupted from above.

Our architecture assigns a UMPRS agent process to each UGV, which means that, in principle, a UGV can elaborate plans both before and during a mission. In the degenerate case, the UGV's UMPRS agent process maintains a single goal of passing the low-level sequence received from above on to the robotic control system. More generally, however, our assumption is that the operator and OWS will *not* elaborate missions into completely executable robotic plans, both because of the centralized computational burden this imposes at the operator level (failing to take advantage of inherent parallelism), and because of the lack of awareness and flexibility that this imposes on the vehicles. Therefore, the model we generally assume is that the operator and OWS elaborate the mission down to the level where coordinated subgoals can be distributed to the vehicles, and it is up to each vehicle to decide how best to accomplish the objectives that it is handed. The vehicles are thus participants in the premission planning phase, elaborating a sequence of subgoals received from the OWS into detailed robotic control commands, in precisely the same way that UMPRS elaborates goals into more detailed steps at the OWS level.

A clear advantage of this strategy is that the UMPRS mechanisms for plan elaboration are inherently responsive to changing circumstances. When deciding what primitive action to take next, a vehicle will consider alternative ways of accomplishing its next goal, taking into account the current context in which it is operating. As context changes, details of how goals are being accomplished will change (different procedures will be retrieved from the library of standard operating procedures), within the more persistent framework of higher-level procedures that are still being adhered to.

The context in which a vehicle operates includes its awareness of the external environment, the internal status, and the activities of other vehicles. Information from all of these sources must be assimilated to classify the current operating conditions (Kluge et al., 1994). In our architecture, this assimilation is done through fusion processes that act on information stored in the Information Assimilation DataBase (IADB). The IADB uses a CODGER-like blackboard (as developed for autonomous land vehicles at CMU (Stentz, 1990)) to store the information about the world used by the UGV agent process. The information is collected from UGV sensors (e.g., camera), from internal monitors (e.g., fuel gauge), and from communications systems (e.g., radio), and combined into symbolic assessments of the situation (e.g., enemy-nearby? is true). To establish context for invoking a procedure, the UMPRS process can query the IADB. To ensure that a procedure remains valid, the UMPRS process can pose a standing query to the IADB, such that the IADB will respond when the conditions of the query are met. And, in some cases, a query from the UMPRS process can trigger the IADB to return goals back to UMPRS. For example, if UMPRS wants to know whether a route is safe, the IADB might trigger observation point planning, essentially saying that it can answer the question if the vehicle carries out a sequence of observations.

Finally, because we generally assume that missions are being carried out by multiple UGVs, a critical component of context is the status of other friendly vehicles. Many cooperative procedures embed communicative actions for maintaining this context, such as in a "bounding overwatch" maneuver where two vehicles leapfrog through an area, taking turns watching over the other and then being watched over in turn, where the turn taking is often synchronized through messages. However, in cases where messages fail to materialize, or where vehicles should maintain radio silence, the vehicles can use observations made with their sensors to draw inferences about the status of other vehicles (including enemy vehicles) (Huber et al., 1994).

Runtime Replanning

Treating individual vehicles and the user interface process at the OWS each as agents provides considerable flexibility not only to premission planning and flexible execution, but also to how vehicles should respond to deviations between expected and actual circumstances during plan execution. In some cases, simple feedback control mechanisms might suffice, such as swerving to avoid an obstacle. In other cases, plans for contingencies might have already been developed, such as fording a stream when a vehicle discovers that a bridge has been demolished.

However, at times more wholesale replanning might be called for. For example, if the vehicles were intended to perform a triangulated reconnaissance to localize some enemy force, the loss of some of the vehicles could jeopardize the mission. A major plan change might be needed, such as distributing the vehicles in radically new locations to approximate the localization effort, or rapidly moving vehicles among multiple points so that each is making several observations, or aborting the initial objective and pursuing a secondary mission goal.

Several replanning strategies exist. Again, at the most degenerate level, sensory data collected by the vehicles can be directly passed to the OWS and simply displayed to the operator, who is responsible for monitoring plan progress and adjusting the detailed robotic plan as needed (that is, when circumstances invalidate the current plans, including contingency plans downloaded to the vehicles). This gives the operator the greatest "hands-on" control, at the cost of requiring the operator to dedicate complete attention to the UGVs.

Because the OWS is an agent, with the associated running UMPRS process and IADB, it can be allowed to perform replanning under some circumstances. In this case, the vehicles return data to the OWS, which the OWS processes (adds to the IADB, where fusion/interpretation processes can act on it). The OWS is internally "executing" the mission plan, following the progress of the vehicles and updating the operator's map, and so can use the IADB updates to check the context of (components of) the mission plan to identify violations of context. These violations, in turn, trigger the OWS to retrieve alternative procedures for accomplishing goals that are valid for the context, elaborating suitable procedures (potentially requesting input from the operator), and sending new instructions down to the vehicles to supersede the vehicles' current sequences of commands. Because the OWS internalizes knowledge of the mission intent and methods for accomplishing that intent, along with knowledge of what might invalidate the premission plan and contingency plans, it can automate recovery replanning in this way.

Of course, to the extent that the UMPRS agent processes on board the vehicles are themselves tasked in terms of objectives and are knowledgeable about alternative procedures, the vehicles can similarly detect plan deviations and replan. In our system, each vehicle can use its own local IADB to update its knowledge about the context. Context changes trigger the retrieval of alternative plans for achieving the vehicle's goals. In addition, because a vehicle maintains top-level goals including keeping other vehicles and the OWS informed of significant context changes, it volunteers some of the contents of its IADB to others, such that their IADBs are updated, triggering further changes either at other vehicles or across the mission at the OWS. In fact, these top-level goals inherently are geared toward coordination, and can trigger cooperative replanning among the vehicles, should they detect that their current plans are outdated and that the OWS is unable (not responding) to oversee their collective replanned activity.

When vehicles perform local replanning, however, it opens the door for uncoordinated activity since their local responses might fail to achieve a synergistic whole. Or worse, their local responses might be conflicting. To avoid this, it is important that the range of replanning choices available to the vehicles be limited enough to avoid conflict, without being so limited as to unnecessarily constrain the vehicles. General constraints, such as so-called "social laws" can serve this purpose (Shoham and Tennenholz, 1994), but often impose constraints for situations that might be impossible given the plans already adopted by the vehicles. An alternative is to allow the vehicles to exchange explicit information about their revised plans, and to search for a minimal set of commitments that assure conflict-free execution (Lee, 1996).

Implementation and Evaluation

The component technologies to build the interacting agents that bridge the gap between the operator interface and the robotic vehicles, as outlined above, have been implemented and selectively deployed. UMPRS has been implemented as the backbone of each of our agents, both for on-board vehicle planning and execution, and for controlling the functionality of the OWS. For mission planning, UMPRS has interfaced to planning tools for formation planning, route planning, and observation point planning. The IADB has been implemented, and its interface with UMPRS has been built. At present, sensor fusion techniques associated with the IADB are for the most part primitive, such that subtle context changes are not recognized, but blatant ones (such as the appearance of an enemy) are captured. The most advanced technique we have implemented for assimilating information has been geared toward plan recognition using belief networks (Huber et al., 1994; Huber, 1996).

Bringing all of these technologies together for an overall evaluation using real UGVs has proven a challenge, primarily because of the inherent limitations, scarcity, and unreliability of current experimental UGV platforms. As a result, we have tested smaller combinations of technologies on various simpler robotic platforms, and a larger combination of technologies within a realistic, widely used simulation system. In what follows, we summarize our observations.

Reactive Execution in an Outdoor Mobile Robot

We have used UMPRS as a controlling agent process for an outdoor electric utility vehicle (Lee et al., 1994). The vehicle navigation system included the YARF road follower (Kluge and Thorpe, 1995) to allow the vehicle to navigate along roads and sidewalks. It was also equipped with the ability to detect a simple class of landmarks (orange traffic cones), and to drive the vehicle off-road to a specified landmark in the environment.

In our experiments the robot's mission was to carry supplies along a road previously scouted by another vehicle. If the earlier vehicle detected an obstacle or blockage along the road it dropped a marker by the side

Figure 3. Road following and marker detection.

of the road. This marker served to warn the following robot that it should turn off the road and head cross-country towards a landmark in order to rendezvous. This was implemented in UMPRS via several plans and their enabling context conditions. The first plan implemented the case where the robot simply traveled all the way down the road, with the context condition that no marker left by the preceding vehicle had been spotted. The second plan implemented the case where the robot turned off the road, located the landmark, and drove the robot cross-country until it encountered the "other" vehicle, with the context condition that a marker left by the preceding vehicle had been spotted. Figure 3 shows a sample scene from one run. The white squares in the horizontal black bars show the sidewalk edge points found by YARF. Additional white squares mark YARF's least squares estimate of the location of the left edge of the sidewalk. The marker cone in the upper right of the image has been detected, and the center of its base marked with a white square.

In our experiments, our unmanned vehicle would start down the road, and could continue to the end of the road if it never saw the cone. When it encounters the cone, however, its UMPRS process sifts through the vehicle goals and the currently enabled plans for them to segue out of the road-scouting activity and into the rendezvous activity. These experiments thus showed the viability of using UMPRS for real-time reactive control of an unmanned vehicle. They also demonstrated the advantage of the procedural planning approach in providing for reactions to events in the environment that are not tied to prespecified locations, in contrast to map-based approaches like the Annotated Map formalism (Gowdy and Thorpe, 1990).

Plan Recognition for Coordinating Indoor Robots

To study the viability of using information assimilation techniques to support the coordinated execution of robotic plans, we implemented our belief-network-based plan recognition techniques as part of the sensing process of an indoor robot. The goal of this robot was to observe the movements of another robot (with a blatant bar-code to be scanned), to infer the likely destination of that other robot, and to move to a complementary destination (typically, the same destination for a rendezvous). In our simulation experiments, we identified characteristics of the plan-recognition problem that impact decisions as to how long the observer should wait (how sure it should be) before acting upon its projection of where the observed agent is headed (Huber and Durfee, 1995). However, realizing these mechanisms in real indoor robots introduced further observational uncertainties that at times would undermine the inferences. Part of the information assimilation process, therefore, involved combining knowledge of relative vehicle movement abilities and models of sensor uncertainty, together with actual observations, to reach better hypotheses about vehicle movements.

Our experiments showed that the robots could indeed use knowledge about possible destinations along with the information assimilation techniques described above to efficiently rendezvous without explicit communication. Thus, the efficacy of information assimilation in support of planning in domains where explicit communication might be dangerous or impossible was established.

Integrated Multivehicle Mission Planning and Coordinated Execution in ModSAF

A true evaluation of our technologies in a military setting would involve four (working) UGVs with reasonable sensing capabilities operating in an environment where some unexpected contingencies would arise—a tall order. To approximate such a situation, we adopted the use of the ModSAF simulation system (Calder et al., 1993) as a powerful, widely used simulator for military purposes.

The ModSAF simulator provides a rich environment to test and evaluate our system. The simulator allows for one to create vehicles from over 200 different platforms ranging from aircraft to land-based vehicles. For our system, we chose to use UGV vehicles. These UGV vehicles were not endowed with all of the sensors and equipment that the real UGV vehicles contained, but they did have enough sensors to perform the kinds of tasks we were interested in. In ModSAF, vehicles are controlled by Finite State Machines, where the user gives the vehicle a sequence of tasks and the vehicle blindly executes them. For our purposes, we needed to have more control over the vehicle so we built into the simulator the ability for the vehicle to received commands from the UMPRS system, along with the ability to run some FSM tasks such as moving to specified points. In that way, we did not have to reimplement methods for directly controlling the vehicles' navigation: UMPRS could employ these methods when sending low-level commands such as go to some point, turn on and off sensors, read sensors and position of the vehicle, and send communications to other vehicles.

The setup of our system is summarized in Fig. 4. Currently, we have four vehicle agents and one OWS agent, running UMPRS. The OWS agent connects to the legacy graphical user interface (GUI) from Hughes STX. The IADB has been integrated in, along with a route planner and an observation point planner (OPP). The plan recognition subsystem for information assimilation is being integrated, but the integration is not complete at the time of this writing. The knowledge possessed by the agents is fairly impoverished: the OWS is knowledgeable about only a few kinds of missions, and acquires information about those missions therefore in mundane ways (without any need to disambiguate among choices). The UGVs have knowledge to achieve basic mobility and perception goals.

We have tested our system in scenarios ranging from 1 to 4 UGVs. A typical scenario is as follows (Fig. 5). An operator wants to scout ahead across enemy lines, such that the scouts eventually take up observation posts overlooking a suspected enemy encampment. This goal is selected from among the (currently few) options known to the mission planner, and the planner retrieves an appropriate procedure for generating the mission through binding map measures to variables, through prompting the operator for more data, and through invoking other planning tools (such as deciding placements of vehicles during a bounding overwatch). The procedure is executed, and the mission plan is formed in terms of a series of subgoals associated with each vehicle. These subgoal sequences are sent to the different UGVs, which run their own UMPRS processes to elaborate these further into specific robotic behaviors (and, along with these, their associated preplanned contingency behaviors). These behaviors are downloaded to ModSAF, and the vehicles begin executing their plans.

In the course of moving to their ultimate destinations, vehicles eventually become aware of an unexpected enemy in a nearby location. The sensory uncertainty captured in ModSAF means that when, during the mission, this happens is non-deterministic. Upon discovering

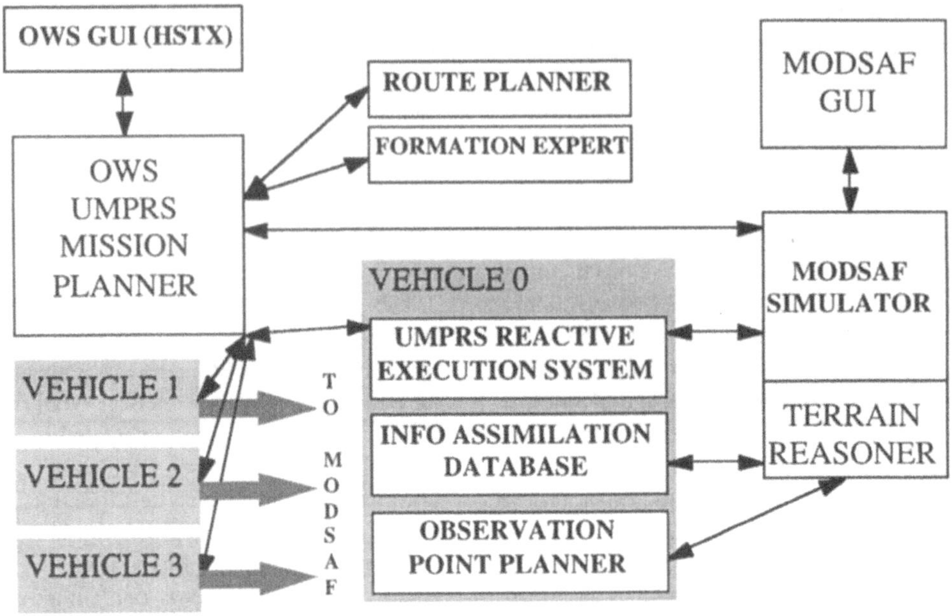

Figure 4. Demonstration setup.

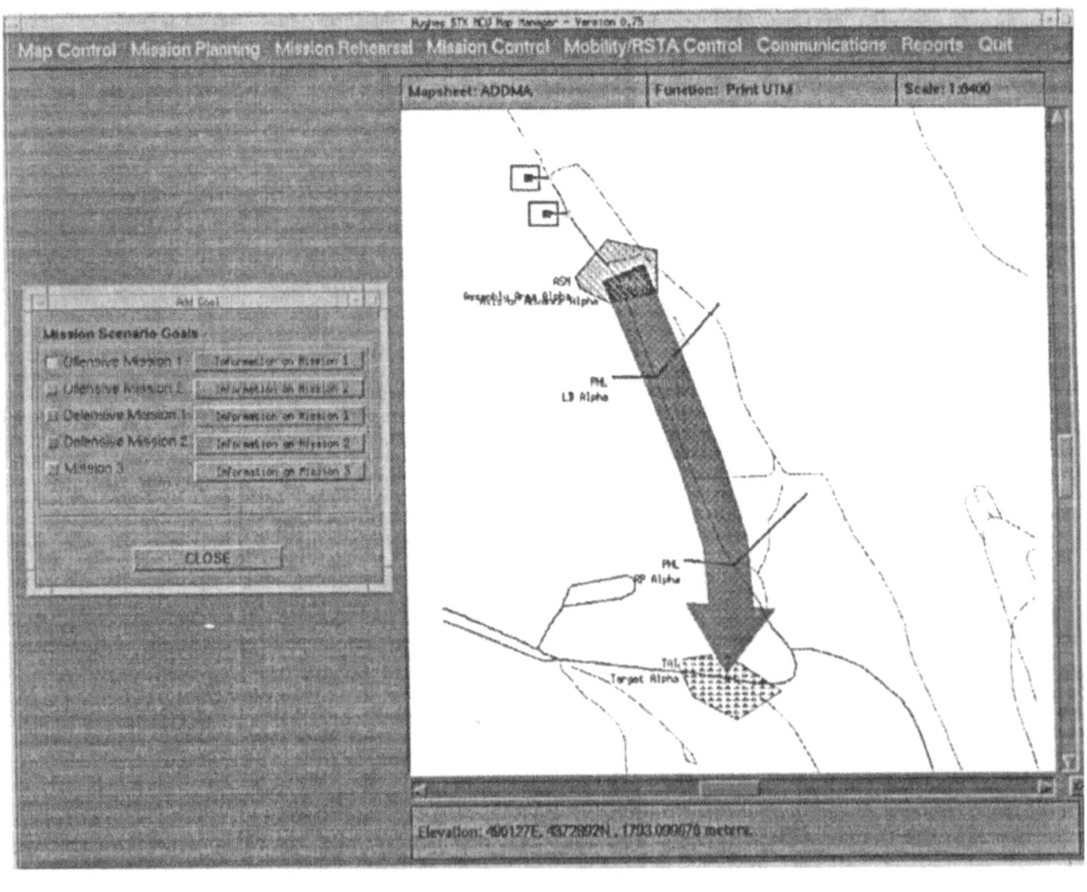

Figure 5. Mission planning display.

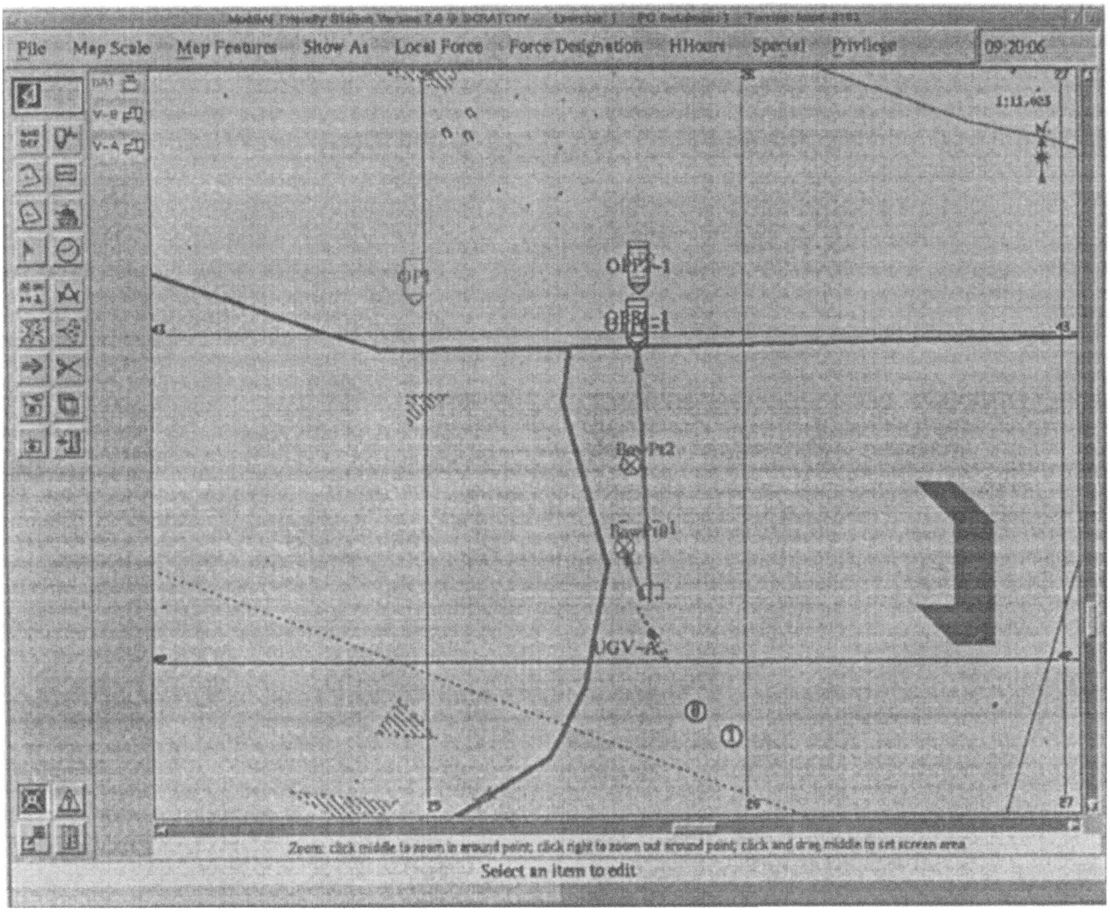

Figure 6. Mission execution in ModSAF.

the enemy, the vehicles alert each other and fall back to previous checkpoints.[1] This is a predefined contingency response, and so is executed very quickly. As they fall back, however, the updated IADB is probed by the UMPRS process on-board the vehicles, and UMPRS on the vehicle that first spotted the enemy feeds the new information into the OPP, which returns new observation points for the vehicle such that it can better isolate the extent of the unexpected enemy presence. The vehicle then leaves its fallback point and carries out this revised mission (Fig. 6).

We have demonstrated this kind of scenario for from one to four vehicles, illustrating the robust accomplishment of critical missions given our techniques. Variations on this theme can have the feedback go up to the IADB at the OWS, and the OWS could plan new observation points for several vehicles rather than having a single vehicle take on the task alone. Or the feedback can go up such that the vehicles wait at their fallback points until the operator explicitly downloads new robotic plans.

The strength of our approach can be appreciated by comparing it to the mission planning as done with the real UGVs during a demonstration led by Lockheed Martin (LM). In the LM demonstration, the development of a mission plan for the three vehicles could take several hours of two operators' time. The operators would micromanage the vehicles during execution. Sometimes they would get calls from the field, requesting that a vehicle be inched forward a little to look over a ridge. Sometimes they would use views from the UGV camera to realize that the UGV was going to the wrong place, and would quickly generate and download a new sequence of robotic actions.

In contrast, by incorporating knowledge into the OWS and UGVs, the demands on operator attention during mission generation and execution are greatly reduced. In cases where the mission template is already known to the OWS, it is only a matter of minutes. And, during execution, more responsiveness is embedded in the automation tools, allowing much more rapid (and less operator-intensive) retasking.

The experiments from the fielded prototype indicated a strong interest on the part of the military personnel for these systems. Even despite the sometimes slow deployment of the UGVs and the intense monitoring costs on the operators, the use of the technology was viewed as a major win because it reduced casualties in a wargame scenario. A major complaint was about the effort required to formulate and monitor UGV missions, however, and it is for this reason that we expect the agent-based technologies that we have developed to have a significant impact on future incarnations of the UGV system. Our demonstrations to military personnel have elicited similar predictions from them as well.

Lessons Learned

To some extent, the lessons learned, as outlined in this paper, touch on social, rather than technical, issues of satisfying operator needs and embracing legacy systems. However, the technical accomplishments of this project provide a flexible means of satisfying these needs, as well as providing the foundation for integrating and coordinating the capabilities of robots, humans, and information systems.

The first lesson that we have learned is that knowledge-based plan execution systems, such as UMPRS, can provide a powerful, tailorable substrate for operator interaction. Systems such as PRS and Soar have already demonstrated ability to control systems in problems including malfunction diagnosis and handling (Ingrand et al., 1992), aircraft sequencing (Rao and Georgeff, 1995) and air combat (Rao et al., 1992). At the level of multivehicle control, our approach is similar to others; we have also demonstrated, however, that the reactive plan execution capabilities can be concurrently employed for operator interaction to provide mixed-initiative definitions of plans and constraints. These capabilities have found other uses beyond UGVs (Durfee et al., 1997).

A related lesson is on the importance of acquiring and propagating intent. Because vehicles might only intermittently be in touch with each other, it is critical that each know enough about the collective goals to forge ahead alone. The robust accomplishment of missions despite communication failures is in no small way due to our strategy of propagating goals and constraints, rather than commands, through the hierarchy.

Of course, properly coordinating the reactive execution of plans requires that the vehicles have sufficient understanding of the evolving plans and goals of others. Plan recognition is thus a critical part of the process of assimilating sensory information. Though we have not gone into the details of the process, our system provides agents with the ability to transform a library of executable plans into a belief-network representation that is especially well-suited to plan recognition (Huber et al., 1994).

Finally, having recognized (or received through communication) the plans that others are pursuing, coordinated execution requires an analysis of the possible outcomes of joint actions, and the establishment of commitments to prevent conflicts. Our experience has shown that there is no single best way to accomplish this in a complex, time-critical domain. Offline commitment to contingency plans, and to constraints of the degree to which plan deviations will be tolerated is one important mechanism. Centralized runtime replanning is another. A hybrid method that we have developed for this project has been to allow runtime examination of agents' adopted reactive plans to establish minimally limiting short-term constraints on reactions (Lee, 1996).

Conclusions

An agent-based approach to bridging the gap between operators and robotic vehicles provides significant advantages by embedding knowledge and initiative within a semi-automated mission planning and execution system. The future objectives of this work include deploying this system for controlling real UGVs in real military settings, and using both such a deployed system and the current simulation system that we have to evaluate the military benefits of using UGVs. Moreover, these technologies apply to the decentralized and semi-autonomous control of many other kinds of systems, and ideas from this work have been applied in other domains including automated ship systems (Durfee et al., 1997).

At a more fundamental level, an open question is the degree to which authority and responsibility should be shared between the operator, the operator's interface agent process (at the OWS), and the agent processes onboard the vehicles. In part, the answer to this question is sociological. However, there are technological factors in modeling the impact of different divisions of autonomy on mission accomplishment time and success rate, and in the degree to which a better representation of the reactive plans (Lee and Durfee, 1994) can allow the division of autonomy to be divided dynamically, that provide exciting opportunities for future research.

Acknowledgments

This is a revised version of a paper presented at the ACM Autonomous Agents '97 Conference. We would like to acknowledge the contributions to this effort made by numerous people who have been part of our team over the years, especially Marc Huber and Jaeho Lee. This work has been supported, in part, by DARPA under contract DAAE07-92-CR012.

Note

1. When plan recognition is in place, a simple test will be that a vehicle that sees the enemy will fall back, and other vehicles will observe that it is moving to an unexpected destination and surmise (using the belief network) that this is strong evidence that it has seen an enemy.

References

Balch, T. and Arkin, R.C. 1995. Motor schema-based formation control for multiagent robot teams. In *Proc. of the First Int. Conf. on Multi-Agent Systems*, pp. 10–16.

Calder, R.B., Smith, J.E., Courtemanche, A.J., Mar, J.M.F., and Ceranowicz, A.Z. 1993. ModSAF behavior simulation and control. In *Proc. of the Third Conference on Computer-Generated Forces and Behavioral Representation*, Orlando, FL, pp. 347–356.

Cook, D., Gmytrasiewicz, P.J., and Holder, L. 1996. Decision-theoretic cooperative sensor planning. *IEEE Transactions on Pattern Analysis and Machine Intelligence*, 18(10):1013–1023.

Durfee, E.H., Huber, M.J., Kurnow, M., and Lee, J. 1997. TAIPE: Tactical assistants for interaction planning and execution. In *Proc. of Autonomous Agents '97*.

Georgeff, M.P. and Lansky, A.L. 1986. Procedural knowledge. In *Proc. of the IEEE*, 74(10):1383–1398.

Gowdy, J. and Thorpe, C. 1990. Annotated maps for autonomous land vehicles. In *Proc. of the 1990 IEEE Conference on Systems, Man, and Cybernetics*, pp. 282–288.

Huber, M.J. 1996. Plan-based plan recognition models for the effective coordination of agents through observation. Ph.D. Dissertation, Univ. of Michigan.

Huber, M.J., Durfee, E.H., and Wellman, M.P. 1994. The automated mapping of plans for plan recognition. In *Proc. of the 1994 Conf. on Uncertainty in Artificial Intelligence*, pp. 344–351.

Huber, M.J. and Durfee, E.H. 1995. Deciding when to commit to action during observation-based coordination. In *Proc. of the First Int. Conf. on Multi-Agent Systems*, pp. 163–170.

Ingrand, F.F., Georgeff, M.P., and Rao, A.S. 1992. An architecture for real-time reasoning and system control. *IEEE Expert*, 7(6):34–44.

Kluge, K., Weymouth, T., and Smith, R. 1994. Information assimilation research at the University of Michigan for the ARPA unmanned ground vehicle project. In *Proc. of SPIE Sensor Fusion VII*.

Kluge, K. and Thorpe, C. 1995. The YARF system for vision-based road following. *Mathematical and Computer Modeling*, 22(4–7):213–233.

Lee, J. 1996. An explicit semantics for coordinated multiagent plan execution. Ph.D. Dissertation, Univ. of Michigan.

Lee, J., Huber, M.J., Durfee, E.H., and Kenny, P.G. 1994. UM-PRS: An implementation of the procedural reasoning system for multirobot applications. In *AIAA/NASA Conf. on Intelligent Robots in Field, Factory, Service, and Space*.

Lee, J. and Durfee, E.H. 1994. Structured circuit semantics for reactive plan execution systems. In *Proc. of the 1994 National Conf. on Artificial Intelligence (AAAI-94)*, pp. 1232–1237.

Rao, A.S., Moreley, M., Selvestrel, M., and Murray, G. 1992. Representation, selection, and execution of team tactics in air combat modelling. In *Proc. of the Fifth Australian Joint Conf. on Artificial Intelligence*, pp. 185–190.

Rao, A.S. and Georgeff, M.P. 1995. BDI agents: From theory to practice. In *Proc. of the First Int. Conf. on MultiAgent Systems*, pp. 312–319.

Shoham, Y. and Tennenholtz, M. 1994. On social laws for artificial agent societies: Off-line design. *Artificial Intelligence*, 72(1–2):231–252.

Stentz, A. 1990. The CODGER system for mobile robot navigation. *Vision and Navigation: The Carnegie Mellon Navlab*, C.E. Thorpe (Ed.), Kluwer Academic Publishers.

Stentz, A. 1995. Optimal and efficient path planning for unknown and dynamic environments. *Int. Journal of Robotics and Automation*, 10(3):89–100.

Tambe, M. 1995. Recursive agent and agent-group tracking in a real-time, dynamic environment. In *Proc. of the First Int. Conf. on MultiAgent Systems*, pp. 368–375.

Tambe, M., Johnson, W.L., Jones, R.M., Koss, F., Laird, J.E., Rosenbloo, P.S., and Schwamb, K. 1995. Intelligent agents for interactive simulation environments, *AI Magazine*, 16(1):15–39.

Edmund H. Durfee is an Associate Professor of Electrical Engineering and Computer Science and holds a joint appointment at the School of Information at the University of Michigan. He received his A.B. degree in Chemistry and Physics from Harvard University in 1980, and his M.S. and Ph.D. degrees from the University of Massachusetts in Computer Science and Engineering in 1984 and 1987, respectively. His research interests are in distributed artificial intelligence, planning, cooperative robotics, and real-time problem solving. He is a 1991 recipient of a Presidential Young Investigator award from the National Science Foundation to support this work, and was an invited speaker at the National Conference on Artificial Intelligence in 1992. He is also an associate editor for IEEE Transactions on Systems, Man, and Cybernetics, and is the program co-chair for the 1998 International Conference on Multi-Agent Systems.

Patrick G. Kenny is a staff researcher at the University of Michigan AI Lab. He received his Bachelor Degree from the University of Minnesota in Computer Science. His interests are in Autonomous Agent Architectures, Intelligent control of Robots and Distributed Minirobotics.

Karl Kluge received his BS degree in Computer Science from Michigan State University in 1985, and his MS degree in Computer Science from Carnegie Mellon in 1988. He received his Ph.D. in Computer Science from Carnegie Mellon in 1993 for his work on the YARF vision-based road following system. He is currently an assistant research scientist in the Electrical Engineering and Computer Science Department at the University of Michigan, although he still roots for Michigan State. His research interests are computer vision, robot navigation, observation planning and sensor fusion, and planning/perception interactions. He is a member of the IEEE.

Autonomous Robots 5, 111–125 (1998)

Learning View Graphs for Robot Navigation

MATTHIAS O. FRANZ, BERNHARD SCHÖLKOPF,**
HANSPETER A. MALLOT AND HEINRICH H. BÜLTHOFF
Max-Planck-Institut für biologische Kybernetik, Spemannstraße 38, 72076 Tübingen, Germany
Matthias.Franz@tuebingen.mpg.de
Bernhard.Scholkopf@tuebingen.mpg.de
Hanspeter.Mallot@tuebingen.mpg.de
Heinrich.Buelthoff@tuebingen.mpg.de

Abstract. We present a purely vision-based scheme for learning a topological representation of an open environment. The system represents selected places by local views of the surrounding scene, and finds traversable paths between them. The set of recorded views and their connections are combined into a graph model of the environment. To navigate between views connected in the graph, we employ a homing strategy inspired by findings of insect ethology. In robot experiments, we demonstrate that complex visual exploration and navigation tasks can thus be performed without using metric information.

Keywords: visual navigation, topological maps, environment modeling, exploration, cognitive maps, mobile robots, omnidirectional sensor

1. Introduction

To survive in unpredictable and sometimes hostile environments animals have developed powerful strategies to find back to their shelter or to a previously visited food source. Successful navigation behaviour can already be achieved using simple reactive mechanisms such as association of landmarks with movements (Wehner et al., 1996) or tracking of environmental features (Collett, 1996). However, for complex navigation tasks extending beyond the current sensory horizon, some form of spatial representation is necessary. Higher vertebrates appear to construct representations—sometimes referred to as *cognitive maps*—which encode spatial relations between relevant locations in their environment (see, O'Keefe and Nadel, 1978; Gallistel, 1990, for reviews).

Under certain conditions, such maps can be acquired visually without any metric information. Humans, for instance, are able to navigate in unknown environments after presentation of sequences of connected views (e.g., O'Neill, 1991). This has led to the concept of a *view graph* as a minimum representation required to explain experimentally observed navigation competences (Schölkopf and Mallot, 1995; Gillner and Mallot, 1997). A view graph is defined as a topological representation consisting of local views and their spatial relations. Depending on the task, these relations can be, e.g., movement decisions connecting the views, or mere adjacencies.

Motivated by the findings of vertebrate ethology, researchers have started to investigate topological representations for robot navigation (e.g., Kuipers and Byun, 1991; Mataric, 1991). These systems rely primarily on local sonar patterns for the identification of places, in combination with metric knowledge derived from compasses or wheel encoders. Bachelder and Waxman (1995) have reported results on a vision-based topological system which uses a neural control architecture and object recognition techniques for landmark detection. In their current implementation, however, the system has to rely on artificially illuminated landmarks and a pre-programmed path during exploration of the

*A preliminary version of this paper was presented at the First International Conference on Autonomous Agents in February 1997.
**Present address: Gesellschaft für Mathematik und Datenverarbeitung (GMD), FIRST, Rudower Chaussee 5, 12489 Berlin, Germany.

environment. For maze-like environments, Schölkopf and Mallot (1995) have shown that learning a graph of views and movement decisions is sufficient to generate various forms of navigation behaviour known from rodents. The scheme has subsequently been implemented in a mobile robot (Mallot et al., 1995).

It is often not clear which of the features of these robotic systems can actually be attributed to the topological representation since metrical and topological knowledge is used together. The same problem arises when interpreting biological navigation experiments since animals usually use various kinds of information depending on the task they have to perform. Therefore, it is interesting from both the robotics and the biological point of view to study systems which are restricted to the use of just one type of knowledge.

The purpose of the present study is to extend the view graph approach from the mazes of Schölkopf and Mallot (1995) to open environments. We will demonstrate that view graphs can be learned autonomously in an environment with complex visual stimuli, without making use of beacons or artificial landmarks. In doing so, we present a navigation system that uses purely topological information based on visual input. By focusing on just one type of information we want to make the contribution of topological knowledge explicit.

In the next section, we describe the experimental setup, followed by an overview of the system's architecture and an account of the required basic mechanisms, namely the procedures for selecting representative views, homing and exploration. We present experimental results in Section 4, and conclude our study by discussing possible extensions of the view graph approach.

2. Experimental Setup

Robot experiments were conducted in an arena sized 118×102 cm. Visual cues were provided by model houses (see Fig. 1). We used a modified Khepera

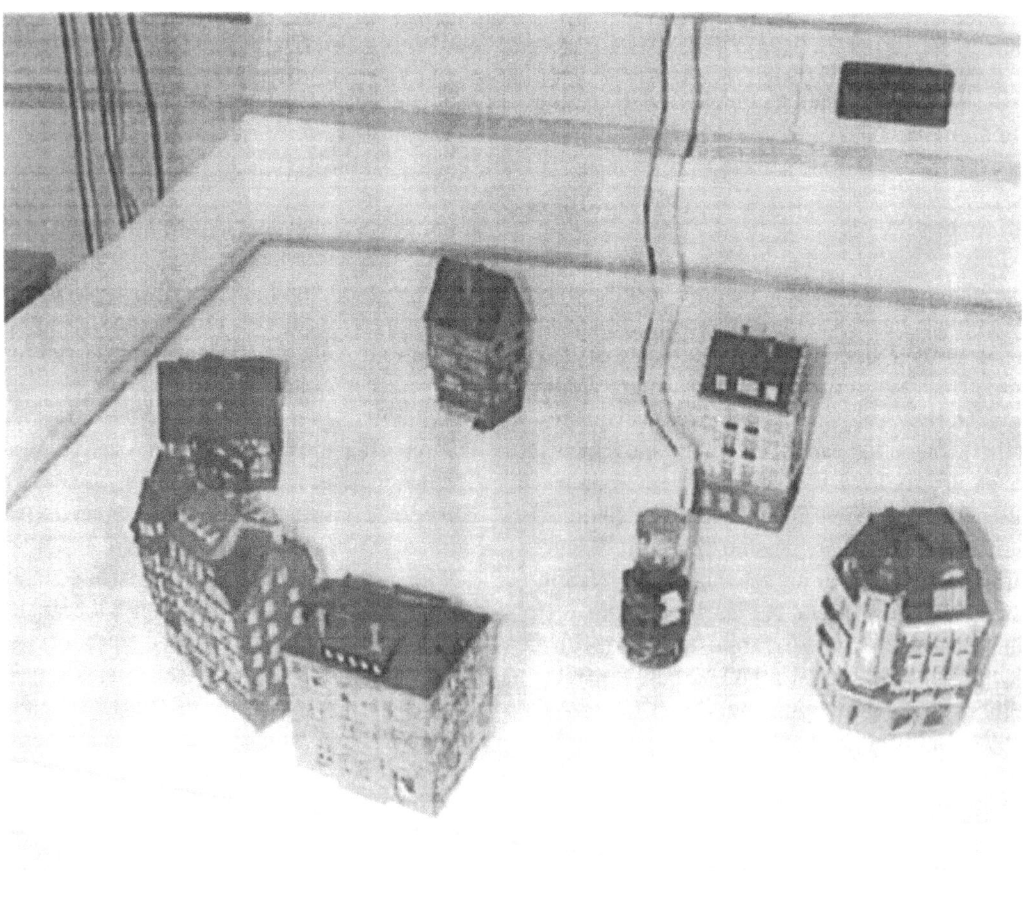

Figure 1. 118×102 cm sized test arena with toy houses and Khepera robot.

Figure 2. Khepera™ robot with camera module and custom made conical mirror, which permits sampling of the environment over 360°, in a range of ±10° about the horizon.

miniature robot (Fig. 2) connected to an SGI Indy workstation via a serial and video transmission cable. Obstacles in a range between 0.5 cm and 4 cm were detected with 8 infrared proximity detectors.

The imaging system on the robot comprises a conical mirror mounted above a small video camera which points up to the center of the cone (Fig. 2). This configuration allows for a 360° horizontal field of view extending from 10° below to 10° above the horizon. A similar imaging technique was used by Chahl and Srinivasan (1996) and Yagi et al. (1995).

The video image was sampled at 25 Hz on four rings along the horizon with a resolution of 4.6° and averaged vertically to provide robustness against inaccuracies in the imaging system and tilt of the robot platform. In a subsequent processing stage, a spatiotemporal Wiener lowpass filter (e.g., Goldman, 1953) was applied to the resulting one-dimensional array. To remove changes in the illumination, the average background component was subtracted and, in a final step, the contrast of the array was enhanced via histogram equalization.

The image data was read by an asynchronous processing thread running at 7 to 12 Hz cycle time. In each cycle, the Khepera's infrared sensors were read out using a serial data link with a maximal transmission rate of 12 Hz. The movement commands calculated from the image and infrared data were transmitted back to the robot via the same serial link.

The Khepera's position was tracked with a camera mounted above the arena. Position and image data were recorded with a time stamp and synchronized offline. Position information was not available to the robot during the experiments. Except for Figs. 10 and 11, all dataplots show results of real world experiments obtained with the described setup.

Computer simulations for Fig. 11 were done in a two-dimensional environment containing triangular objects of random size and shading (cf., Fig. 10). Views were computed with standard ray-tracing techniques. The simulated infrared sensors provided distance values corrupted by Gaussian noise. The control architecture and the interface to the simulated motors and sensors were identical to the ones used in the real-world experiments (Franz et al., 1997).

3. Learning View Graphs

3.1. Discrete Representation of Continuous Space

In view-based navigation tasks, visual information is used to guide an agent through space. The reason why this is feasible at all, is the fact that there is a continuous mapping between position space (x- and y-coordinates, possibly supplemented by gaze directions) and the space of all possible views: for each spatial position, a certain view is perceived, and this view changes continuously as the observer moves around in space[1]. Unfortunately, this mapping can be singular, as identical views can occur at different spatial locations, i.e., there is no guarantee for the existence of a global coordinate system on the manifold of all possible views (cf., Fig. 3). In principle, this problem can be dealt with using context information: In points with identical views, we can use views from nearby spatial positions to disambiguate between them.

Complete knowledge of this manifold would be very useful to determine one's spatial position. Memory and computation requirements, however, prohibit storing everything. Moreover, if we are not interested in determining positions at arbitrary times but rather in carrying out specific navigation tasks, as for instance path

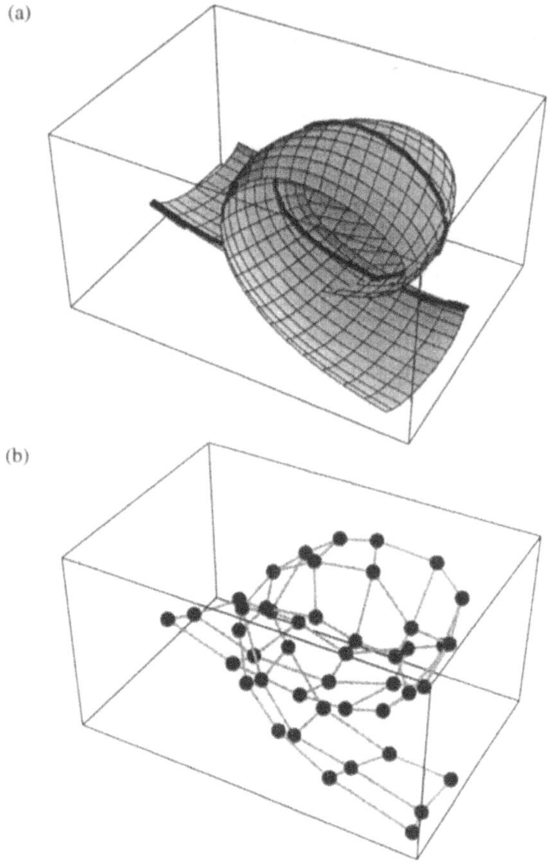

Figure 3. (a) Sketch of a view manifold, consisting of all views that can be seen by a continuously moving observer. The manifold is embedded in a Euclidean space whose dimensionality is the number of camera pixels. The actual structure of the manifold is much more complicated, with holes caused by obstacles, and a tubular structure due to the possibility to take snapshots at all orientations between 0° and 360°. The self-intersection leads to a singularity of the coordinate system inherited from position space: if one moves along the path (black line), the same view occurs twice at different spatial locations. (b) Example of a view graph representation of the view manifold.

planning, this is not at all necessary. In that case, it is sufficient to store views which allow the description of relevant paths. This leads to a less detailed representation of the view manifold, namely by a graph consisting of representative views and connections between them.

In discretized environments like mazes, there is a canonical set of views to store: since no movement decisions need to be taken while traversing corridors, the views necessary to support path planning are solely those at junctions. As open environments do not impose a structure on the view graph, we have to select a set of views which are representative for the manifold (in the following referred to as *snapshots*), and to find connections between them. Since the connecting paths between the snapshots are not explicitly coded in the view graph, we have to provide a homing method which allows us to find connected views from a given start view.

In the following sections, we introduce a system that is able to solve these tasks. The vertices of the acquired view graph are panoramic views of the environment, and its edges are connections between views that can be traversed using a visual homing procedure. In contrast to our previous implementation (Schölkopf and Mallot, 1995), this homing procedure allows the system to approach a location from any direction such that the graph edges denote mere adjacency relations without any directional labelling. The resulting view graph does not contain any explicit metric information.

3.2. A Minimalistic System for Learning a View Graph

The overall architecture of the system is shown in Fig. 4. Here, we discuss the basic building blocks, the details are described in the following sections.

View Classifier. As we mentioned above, a crucial component of any graph learning scheme is the selection of vertices. The graph vertices have to be distinguishable, because otherwise the representation

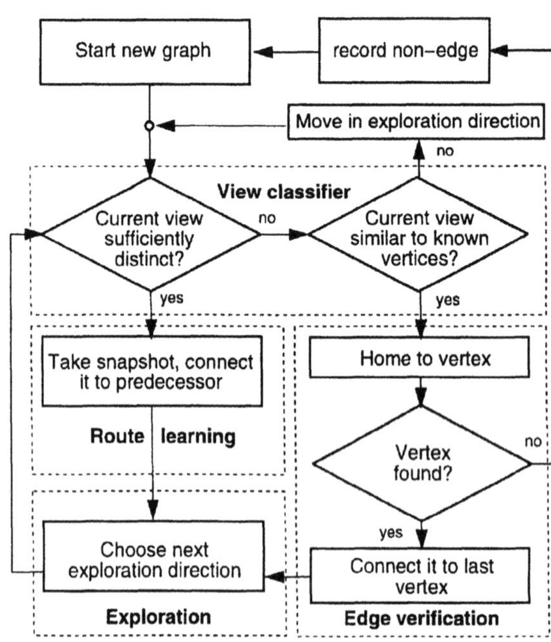

Figure 4. Block diagram of the graph learning algorithm.

could not be used for finding a specific location. Since we confine our system to use only visual information, we must guarantee that the recorded views are sufficiently distinct. This can be done by a classifier which detects whether the incoming views can be distinguished from the already stored vertices of the view graph. If this condition is fulfilled, the system takes a new snapshot and adds it to the graph. The classifier is described in Section 3.3.

Route Learning. In our system, a new snapshot is automatically connected to the previously recorded vertex of the view graph. Thus, the systems records chains of snapshots, or routes. These routes can be used to find a way back to the start position by homing to each intermediate snapshot in inverted order. We describe the homing procedure in Section 3.4. The area from which a specific snapshot can be reached by the homing procedure is called its *catchment area.* The view classifier has to make sure that every snapshot can be reached from its neighbour, i.e., all vertices of the view graph have to be in the catchment areas of their adjacent vertices.

Choice of Exploration Direction. When the system has taken a new snapshot, a new exploration direction must be chosen. This choice primarily affects the *exploration strategy* of our system, since it determines how thoroughly an area is explored and how fast the explored area grows. In the Section 3.5, we describe several local exploration strategies used in our system.

Edge Verification. The route learning procedure described above has no way of forming new edges to previously visited views, i.e., the resulting graphs will be mere chains. By adding the following behaviour we can get nontrivial graphs: During exploration, the system constantly checks whether the current view becomes similar to the already recorded snapshots. This again is a view classification task which can be solved by the same classifier as used for the selection of the snapshots (cf., Section 3.3). In a second step, the system checks whether the detected snapshot is not yet connected to the vertex from which the current exploration step started. Whenever these conditions hold, the system tries to home to the snapshot. If successful, an edge connecting the two vertices is included, and the exploration continues from the detected snapshot. In cases where the robot gets lost or bumps into obstacles, we record a "non-edge" between both vertices thus preventing the failed action from being repeated. Before starting to home, the verification procedure always checks whether a "non-edge" for this action has already been recorded. After a failed verification, we start a new graph, which will typically get connected to the old one in due course by the edge verification procedure.

If an already connected view is encountered during an exploration step the system homes to it as well (not shown in Fig. 4). This procedure does not produce additional knowledge, but has the effect that edges intersecting previously stored edges are less likely to be recorded. Edge verification could in principle also be used for the edges learned by the route learning procedure. For reasons of excessive exploration times, we did not resort to this more cautious strategy.

Arbitration and Obstacle Avoidance. As the focus of this paper is on navigation, we did not use any sophisticated obstacle avoidance systems. During exploration, the infrared sensors of the robot were continuously checked for the presence of nearby objects. If obstacles were detected at distances larger than ca., 1 cm, the robot tried to turn away without slowing down. Smaller distances were interpreted as collisions causing the robot to back up and turn away from the obstacle. Both behaviours and the graph learning system of Fig. 4 were combined into a subsumption architecture (Brooks, 1986), where the collision-induced escape behaviour had highest, the graph learning procedure lowest priority. This simple architecture proved to be sufficient to guide the robot through the environment shown in Fig. 1.

The robot is not allowed to take snapshots as long as the obstacle avoidance system is active. The resulting graph structure tends to concentrate in the open space between obstacles. This feature makes the navigation system more effective, as the visual input changes very rapidly near objects. Exploration of these areas would require a large number of snapshots which, in complex natural environments, would ultimately lead to a fractal graph structure near objects.

3.3. *View Classifier*

In our robot implementation, the vertices of the view graph are identified with snapshots of the surrounding panorama, namely one-dimensional 360° records of the grey values at the horizon. Omnidirectional views have the advantage that one view alone can encode a location without requiring the computationally expensive merging of several views to one place representation. In addition, this feature provides robustness against changes in the image affecting only parts of the panorama, e.g., moving objects. Global changes in the lighting conditions are removed by the image processing (cf., Section 2) such that the view manifold in our setup is relatively stable over time.

Ideally, the set of snapshots taken to represent the view manifold should satisfy three criteria: First, the views should be distinguishable. In purely graph-based maps, this is the only way to guarantee that specific vertices can be navigated to. This can be achieved by incorporating only distinct views into the graph. Second, a large proportion of the view manifold should be covered with a small number of vertices to keep processing requirements small. Third, the spatial distance of neighbouring views should be small enough to allow reliable homing between them.

As we confine our system to use only visual input, the selection of the snapshots must be based on the current view and the stored snapshots. The above criteria can be satisfied by measuring the degree of similarity between views: Dissimilar views are distinguishable by definition while being distant on the view manifold, and similar views often are spatially close.

For measuring similarity, we take a minimalistic approach by using the maximal pixel-wise crosscorrelation Φ as a measure of similarity. This is equivalent to the dot product of two view vectors a_i, b_i, after first rotating one of them such as to maximize the overlap with the other one:

$$\Phi = \max_i \sum_j a_j b_{j-i}. \quad (1)$$

Whenever a threshold of the image distance to all stored snapshots is exceeded by the current view, a new snapshot is taken. The threshold classifier thus adapts the spacing between the snapshots to the rate of change in the optical input. Areas which have to be covered by a denser net of snapshots, due to a rapid change of views, are also explored more thoroughly. The threshold is chosen to ensure that the snapshots are both distinguishable and close enough to allow safe navigation between them. Larger thresholds reduce the number of recordable vertices and increase the spacing between them. Smaller thresholds lead to a denser net of snapshots, but increase the danger of incorporating false edges due to the confusion of vertices. Note, that the spatial position of the snapshots is influenced to a large degree by the position of the first snapshot since all subsequent snapshots are selected according to the internal distinctiveness criterion. For varying start positions, the recorded view graph will always look different.

Clearly, a threshold classifier can also be used to detect whether the current view becomes similar to one of the already recorded snapshots. If the image distance to a snapshot falls below the threshold, the robot starts its edge verification procedure (cf., Section 3.2) and tries to home to the snapshot. In our system, we use the same classifier for both tasks. A suitable threshold was determined experimentally (cf., Section 4.1).

3.4. *Navigating between Places: View-Based Homing*

In order to travel between the vertices of the view graph, we need a visual homing method. Since the location of a vertex is only encoded in the recorded view, we have to deduce the driving direction from a comparison of the current view to the goal view. After the robot has moved away from the goal, the images of the surrounding landmarks in the current view are displaced from their former image positions in the goal view. If the robot moves so as to reduce these displacements it will finally find back to the goal, where current view and snapshot match. The displacement field has a focus of contraction in the goal direction (cf., Fig. 5). Driving into the direction of this focus most quickly reduces the image displacements.

A number of experiments have shown that invertebrates such as bees or ants are able to pinpoint a location defined by an array of nearby landmarks (see, Collett, 1992, for a review). Apparently, these insects search for their goal at places where the retinal image forms the best match to a memorized snapshot. Cartwright and Collett (1983) have put forward the hypothesis that bees might be able to actively extract the goal direction by a homing mechanism as described above.

In order to apply the idea of Cartwright and Collett (1983) to robotic homing tasks, two basic problems have to be solved:

1. Correspondences between image points in the snapshot and in the current view must be established to detect displacements between them.
2. If a visually navigating agent has no direct access to the actual distance of the surrounding landmarks, this lack of knowledge must be compensated by some additional assumption about the distance distribution of possible landmarks in the environment.

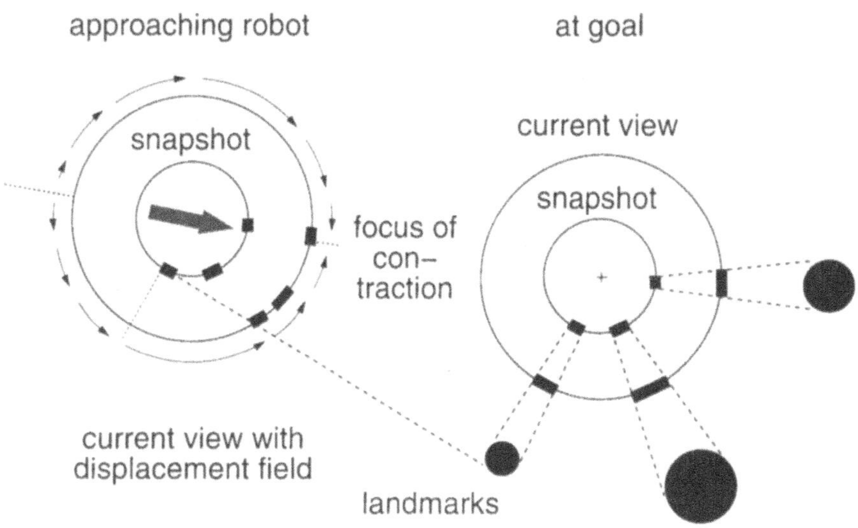

Figure 5. A robot with omnidirectional sensor uses a snapshot taken during a previous visit to find back to the goal position. After moving away from the goal, the images of the surrounding landmarks are displaced to new image positions, as denoted by the arrows at the current view. If the robot moves such that it reduces the displacements it will finally find back to the goal (after Cartwright and Collett, 1983).

In our approach, we assume that all visible landmarks have approximately the same distance from the location of the snapshot. The resulting displacement fields $\delta(\theta)$ have a very simple structure (Franz et al., 1997)

$$\delta(\theta) = \arctan\left(\frac{\nu\sin(\theta-\alpha)}{1-\nu\cos(\theta-\alpha)}\right) - \psi, \qquad (2)$$

where θ denotes the image position of a landmark in the snapshot, α the direction in which the robot has moved away from the goal, ψ the change in the sensor orientation, and ν the ratio between the distance from the goal and the average landmark distance. Since these displacement fields depend only on three free parameters ψ, α and ν, they can be used as matched filters: From the parameters of the best match to the actual displacement field, we compute the driving direction given by $\alpha + \pi$. These steps are repeated until the current view becomes identical to the snapshot.

```
REPEAT {
  FOR all values of ψ, α, ν DO {
    compute displacement field from Eq.(2)
    distort snapshot with displ.field
    compute image distance to current view
  }
  select parameter set with min.distance
  drive in direction α + π.
}
UNTIL image distance to snapshot is 0
```

The matching of three parameters requires relatively small computational resources compared to other methods for image matching. On an SGI Indy workstation, the calculation of a home vector from 78-dimensional views took less than 40 ms which results in smooth trajectories to the home position. It can be shown mathematically (Franz et al., 1997) that the goal can be approached with arbitrary accuracy even though the differences in the distances to the individual landmarks are neglected, and that each snapshot is surrounded by a non-empty catchment area. In practice, the accuracy depends mainly on the quantization and sensor noise of the detector ring, since a displacement can only be detected if it generates sufficient change in the detector signal. In our experimental setup, this was usually the case at distances from the goal in the range of 1 to 3 cm, depending on the distances of the surrounding landmarks. The size of the catchment area for a single snapshot is mainly determined by the layout of the environment. In our toy house arena, maximum homing distances of 45 cm were achieved. The success rate was 95% for homing distances smaller than 15 cm, and dropped to 50% in the range of 20 to 25 cm.

The displacements in panoramic views after observer movements have been used in several robotic systems for homing tasks. For example, Hong et al. (1991) identified and matched image features in images with constant orientation. They used this scheme to successfully guide a mobile robot along a corridor. In Röfer's system (1995), a Kohonen network was used to learn the correspondence between snapshot and current view.

3.5. Local Exploration Strategies for Graph Learning

The exploration strategies used by our robot have been motivated by the principle of *maximizing knowledge gain* (Thrun, 1995). As we have not formalized any notion of knowledge, this principle was used as a qualitative guideline. In our context, knowledge gain is possible, for instance, through the recording of new edges and new snapshots. In the following, we describe several exploration strategies, which concern primarily the choice of the next direction to explore after a snapshot has been taken, or after an existing vertex has been reached (cf., Section 3.2).

Exploration Direction during Route Learning. The simplest conceivable rule is to choose a random direction and then to go straight until the next snapshot is taken. The resulting Brownian motion pattern has the advantage that eventually every accessible point of the environment will be explored without the danger that the exploring agent is caught in an infinite loop. Good results can also be achieved if one uses a fixed turning angle. Using smaller angles, distant areas are reached faster, whereas angles closer to π lead to a more thorough exploration of the local neighbourhood. For the experiments presented here, we used a fixed turning angle of 90° to the left.

Exploration of the Largest Open Angle. Our navigation scheme is designed such that all vertices of the view graph remain in the catchment areas of their respective neighbours. This property can be used to choose the next exploration direction, if a vertex has already more than one edge: The system determines the directions to all neighbouring vertices using the homing procedure, and directs the next exploration step to the largest open angle between them (cf., Fig. 6). Alternatively, one could use information about neighbouring vertices, such as their connectivity or similarity. For example, exploring areas where neighbouring views are connected to each other would be more likely to lead to possibly undesired edge intersections.

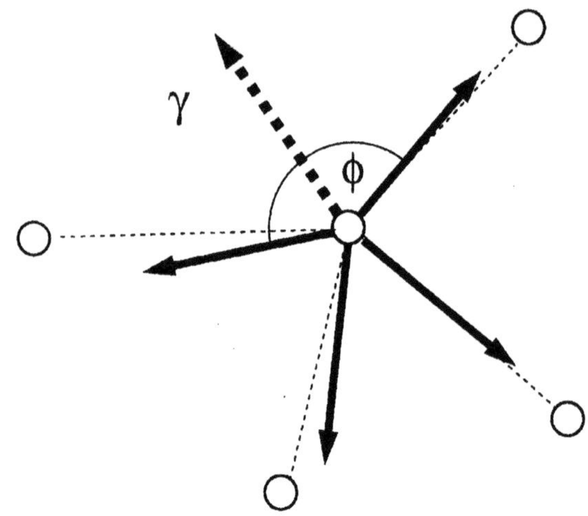

Figure 6. The robot estimates the directions to all connected vertices (black arrows). The largest open angle is denoted by Φ. The next exploration step will be directed along γ into the middle of Φ.

Limiting the Connectivity of Vertices. The exploration can be made more effective by limiting the number of edges a vertex can have. If a vertex reaches the maximum number of edges the robot moves on to less connected neighbours instead of starting a new exploration step. Similarly, the robot determines the directions to all neighbours as described above. If the largest open angle between the directions is smaller than a preset value, the system moves on to other vertices. Using this strategy, exploration tends to spread out to less explored vertices and areas.

Together, these strategies form the routine for the choice of the next exploration direction in Fig. 4:

```
Choose next exploration direction:
n := count number of edges
FOR all connected views DO
  compute directions to connected views
φ := largest open angle
γ := vector pointing into the middle of φ
IF n > max edges OR φ < min angle THEN
  move to least connected neighbour
IF n < 2 THEN
  exploration direction := fixed angle
ELSE
  exploration direction := γ
```

In this study, we were only interested in evaluating the performance of local rules, but the approach can easily be extended to include global rules such as searching the graph for less explored vertices, or deleting unnecessary edges.

4. Results

4.1. View Classifier Threshold

In order to find the threshold of the image distance for the view classifier, we made the following experiment (Fig. 7): During a test run, the robot covered the entire free space of the arena with snapshots spaced at most 2 cm apart. From these snapshots, view pairs were selected randomly and their image distance was computed. The range of image distances was divided into 6 bins. For each bin, homing runs for 20 different view pairs were performed. A run was counted as a success if the robot reached a 1 cm radius around the home position in less than 30 s.

Since the classifier is used for two different tasks, 20 view pairs were generated for each of the following two categories: (1) Pairs connected by a direct line of sight are relevant for the selection of snapshots, since there must be traversable space between them during exploration. (2) For edge verification (see Section 3.2), no restrictions on the set of possible views can be made. Figure 7 shows the success rate for both categories. We have chosen the threshold value Θ (dashed line) such that vertices connected by a direct line of sight can be reached in 90% of all cases (light grey bars). For the general case (dark bars), where the path may be blocked by obstacles, at least 70% of all vertices with image distance below Θ can still be found by the homing procedure.

As an example, we have computed the image distance map to a snapshot at the center of the arena for the view dataset mentioned above (Fig. 8). Regions with image distances below Θ are surrounded by white contours. If the robot starts at the center, the next snapshot would be taken after the white contour is crossed. This leads to a spacing of the snapshots between 5 and 15 cm, depending on the variability of the visual input. The number of snapshots that can be distinguished using this classifier usually falls in a range between 20 and 40, depending on the start position. This is due to the low resolution and contrast of the mirror optic of the robot and the occurrence of similar views in different parts of the arena.

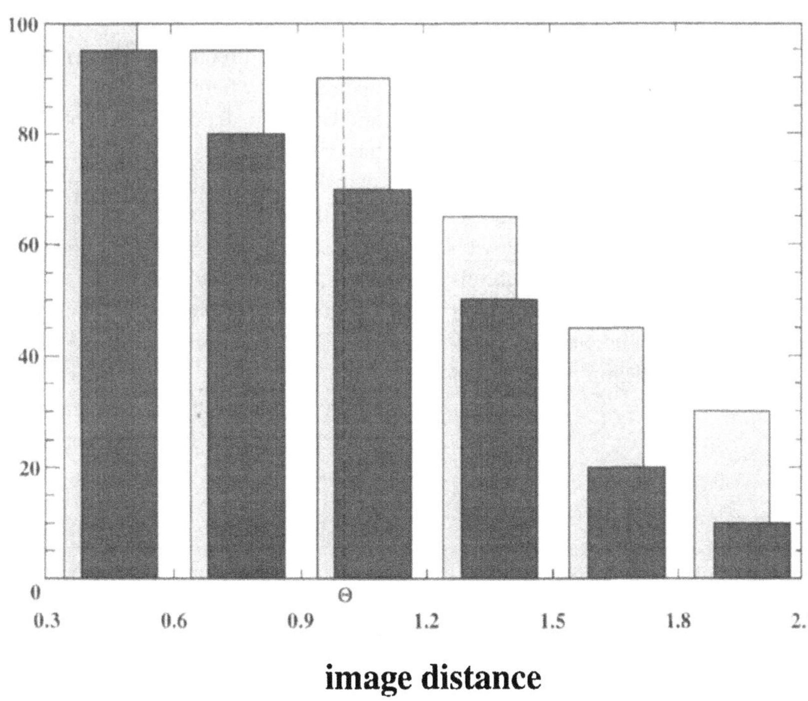

Figure 7. Success rates (in %) for travelling between views with various image distances. In each bin, 20 pairs were tested. Light grey bars are results for pairs connected by a direct line of sight, dark bars for general pairs. The dashed line marks the threshold Θ of the classifier; image distances are measured in multiples of Θ.

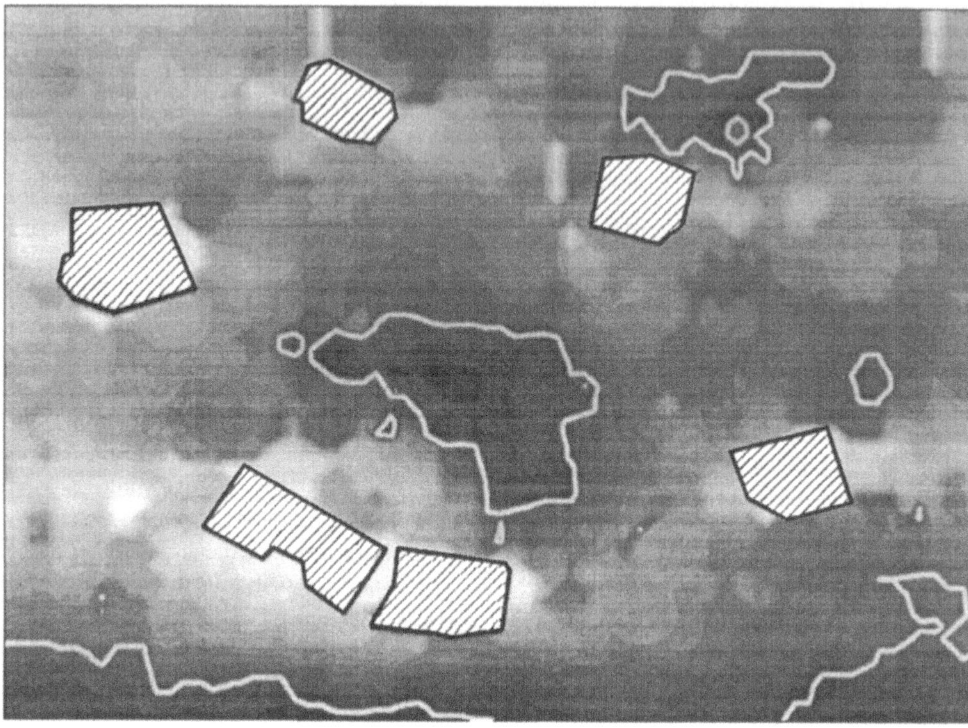

Figure 8. Map of image distances to a snapshot at the center of the arena. Darker regions represent lower image distances. White contours enclose regions with an image distance below the threshold Θ. Hashed regions mark the shape and position of the toy houses in Fig. 1.

If the threshold is used to detect similarity, there are several regions with false positives. In these regions, the edge verification procedure would try to find new edges to the central view, but fail to find it since the view never becomes similar enough to the snapshot. Although our setup allows only to record 20–40 snapshots, the visual input is sufficiently rich to prevent false edges due to identical views in different locations. Clearly, in some environments, e.g., in corridors, identical views may be more likely to occur, and thus would require additional information to disambiguate the locations.

4.2. *Graph Learning*

Figure 9 shows some features of the graph learning scheme. The trajectories were recorded using the tracking system described above. The robot starts by recording a chain of vertices 1 to 4, turning at a fixed angle of 90° after each snapshot. After leaving vertex 4, the current view becomes similar to vertex 1. The edge verification procedure establishes a new edge between vertex 1 and vertex 4 by homing to vertex 1. The next exploration direction is chosen to fall into the largest open angle, where the robot collides with an obstacle and backs up. It continues exploring until the momentary view becomes sufficiently different from all stored snapshots and starts a new graph by recording vertex 5. This time, the proximity of vertex 2 is detected, and a new edge between vertex 2 and 5 is established, connecting the two subgraphs.

In order to illustrate the accumulation of knowledge over time, we used a simulated two-dimensional environment (cf., Fig. 10) as described in Section 2. This environment provided sufficiently rich visual input to record more than 100 snapshots in the view graph and allowed for long exploration times. The simulated world can be traversed in 30 time units, thus in the overall time period depicted in Fig. 11, the arena was crossed approximately 50 times. We first observe an increase in the number of both graph vertices and edges. However, after some time almost no new snapshots are taken—the view manifold has been sufficiently densely sampled, while the verification procedure still adds new edges to the graph.

Figure 12 shows two examples of view graphs G_1 and G_2 recorded by the Khepera robot in the toy house

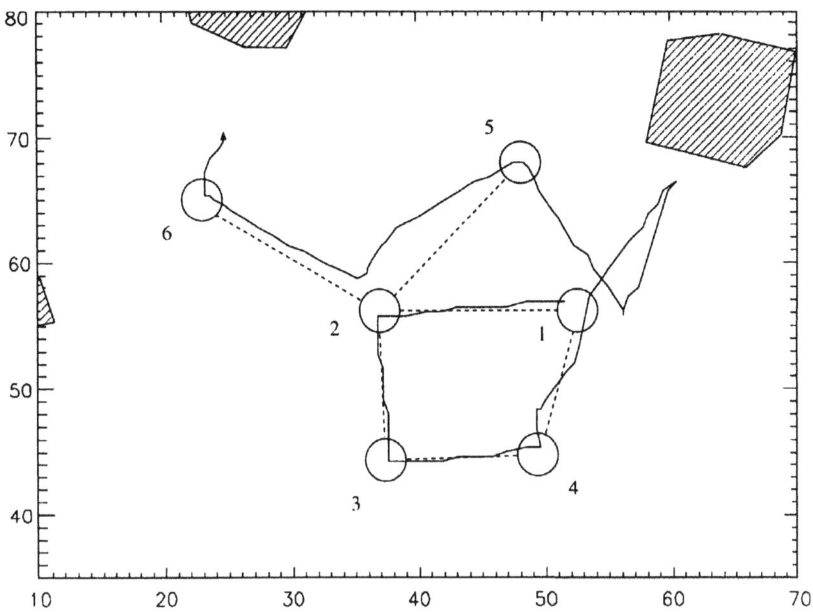

Figure 9. Robot trajectory recorded by the tracking system during an exploration run. Circles denote locations of snapshots, dotted lines recorded edges.

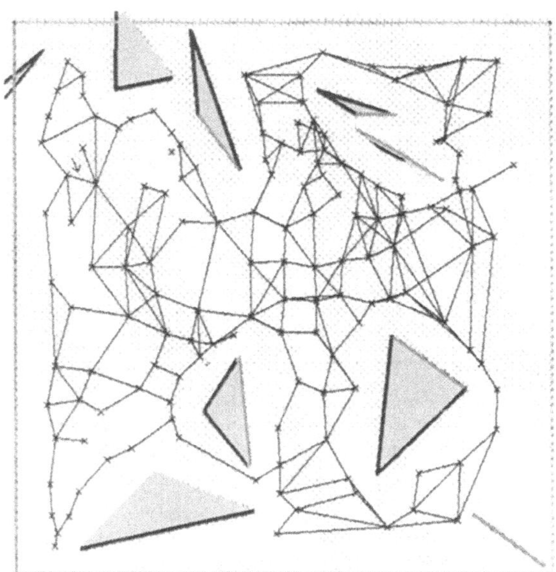

Figure 10. Simulated two-dimensional world with triangular objects of random size and shading. The depicted graph shows the location of snapshots and recorded edges between them.

environment with different start positions. G_1 contains 35 vertices and 46 edges, G_2 21 vertices and 32 edges. After exploration (G_1 75 min, G_2 60 min) all unconnected vertices (6 in G_1, 4 in G_2) were deleted from the graph. 89% of the edges in G_1 and 97% in G_2 could be reproduced in a subsequent homing experiment. Note, that unreproducible edges do not render the graph useless for navigation. Since the threshold of the classifier is chosen such that the vertices remain in the catchment areas of their neighbours, the system does not lose orientation if a particular vertex cannot be found.

Both graphs cover more than half of the open space in the arena. The connectivity of the graphs reflects the topological relations of the environment. As discussed above, the system records only snapshots which are sufficiently distinguishable. This prevents the robot from taking snapshots in the regions outside the areas covered by the graph because there the view is too similar to the already recorded snapshots. Nevertheless, the visual input provided by our setup is sufficiently rich that no exactly identical views occured in the arena such that no false links were recorded.

4.3. *Examples of View Graphs*

Once the graph has been learned, one can generate a path to a goal by search algorithms, e.g., as described by Schölkopf and Mallot (1995), and then sequentially navigate along this path by homing. The use of the view graph for global navigation tasks is illustrated by the sample trajectory in G_1 (Fig. 12). The robot traverses

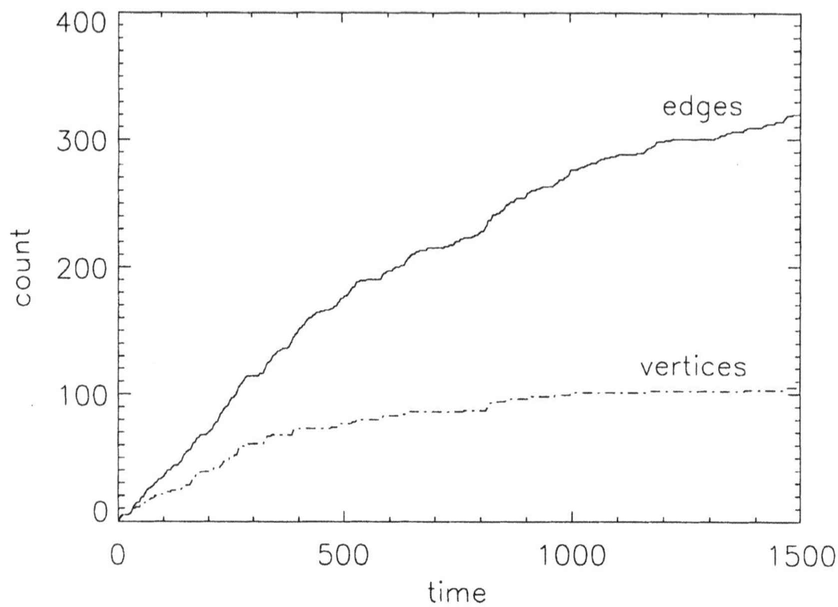

Figure 11. Number of graph vertices and edges as a function of exploration time during a simulated exploration. The number of edges continues increasing when the number of vertices has saturated.

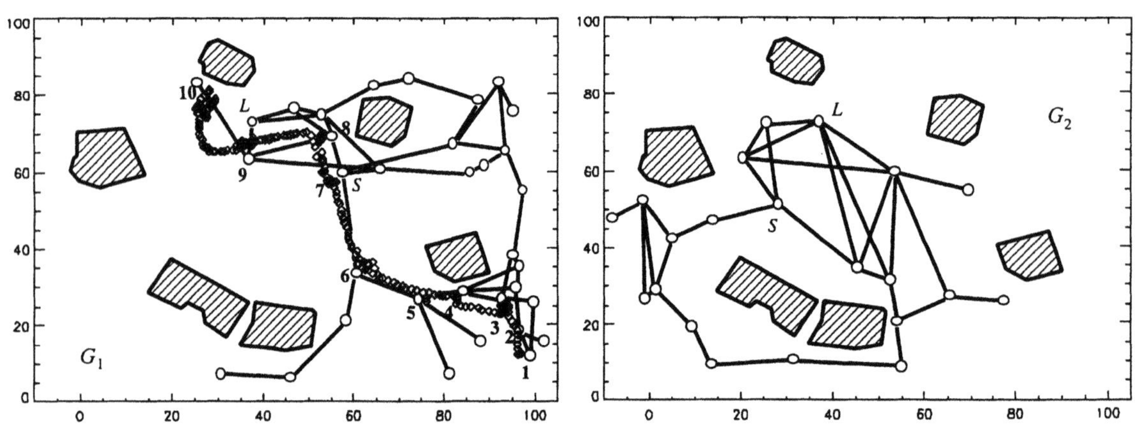

Figure 12. Two view graphs recorded by the Khepera robot with different start positions S. Circles denote locations of snapshots, lines recorded edges between them. The sample trajectory (diamonds) in G_1 started at vertex 1. Vertex L is a possible linking place between G_1 and G_2 (see discussion).

a chain of 10 vertices, thus connecting regions which have no visual overlap.

The final analysis sheds some light on the relationship between topological (i.e., graph-based) and metrical maps. For all view pairs in the graph G_1, we computed both the graph distance (or *combinatorial distance*) and the Euclidean distance in space. It turned out that in our experimental setup, the two distance measures are strongly correlated (Fig. 13). This means that even though our system has not acquired explicit metrical information during exploration (no distances or angles were recorded), the resulting topological map does nevertheless contain some information about metric distances. This is due to the properties of the graph learning system. If two vertices are spatially far apart, any possible edge connecting them would probably intersect other edges, which our system tries to avoid (cf., Section 3.2). If two vertices are close, the edge verification procedure is likely to connect them in the graph.

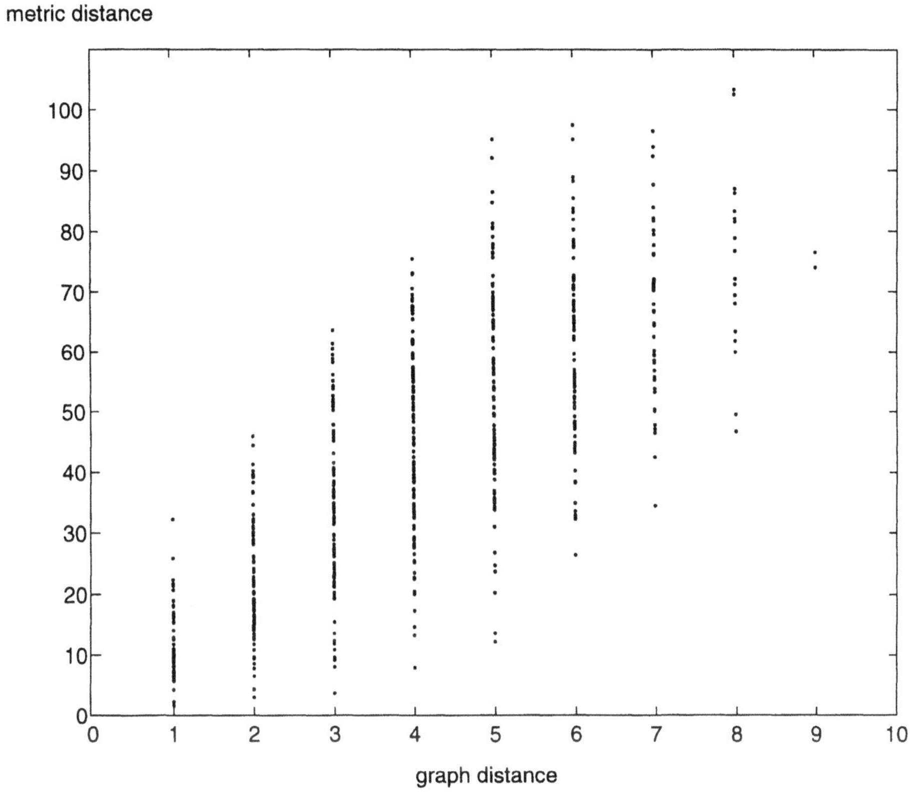

Figure 13. Scatterplot of metric distance vs. graph distance for all connected view pairs of the graph G_1 (Fig. 12). The correlation indicates that graph distance contains information on metric distance.

5. Discussion

In this study, we presented a system which is able to acquire a graph representation of an open environment using only visual information. The purely topological approach relies on the availability of a homing mechanism to reach neighbouring vertices and a simple threshold classifier for selecting snapshots. The robot implementation demonstrates that complex exploration and navigation tasks can be performed without resorting to metric information.

Our experiments have also shown a principal limitation of systems that have no access to metric information during exploration: Only areas providing non-ambiguous information can be mapped reliably. Although the 78-dimensional views used in our system are able to encode more places than, e.g., the relatively low-dimensional signatures of ultrasonic sensor rings, the range of the system could probably be further improved by using views with higher resolution and contrast in combination with more sophisticated classifiers. In addition, views could be made more distinct by considering also their context in the graph, as proposed by Kuipers and Byun (1991) for place identification. Similarly, the neural architecture of Schölkopf and Mallot (1995) utilizes lateral weights to bias view recognition by topological context.

A simple solution to the problem of ambiguous visual input is to use distances and directions for disambiguation during learning. Once the topological representation is learned, no metric information is needed for navigation tasks since the vertices are uniquely defined by their context. Piaget and Inhelder (1967) proposed a similar idea in their theory of early spatial knowledge: While children do not appear to memorize explicit metric information, they seem to use it when they learn to orient themselves in their environment.

Note that disambiguation can also be performed without using metric information by including local graphs such as G_1 and G_2 in Fig. 12 in a collection of linked graphs. Common vertices between subgraphs have to be marked as *linking places* to allow transitions between them (Poucet, 1993). In our example (Fig. 12), vertex L is common both to G_1 and G_2

and could be used to switch from one subgraph to the other.

Using a purely topological representation, our system is necessarily confined to the known path segments coded in the graph. Although it is able to detect neighbouring unconnected vertices, there is no simple way to find novel paths over terrain not contained in the catchment areas of recorded views. However, our experiments have shown that our simple topological representation contains implicit metrical knowledge which might be used to accomplish tasks ususally attributed to a metrical representation. This has implications for the interpretation of experimental results: If an animal can be shown to utilize metrical information, one cannot directly conclude that it was acquired explicitly during exploration.

Several information sources can be integrated into a common graph representation, with vertices containing information about different sensory input and internal states. Lieblich and Arbib (1982) propose that animals use a graph where vertices correspond to recognizable situations. The same idea was also used in the robot implementation of Mataric (1991) where vertices are combinations of robot motions with compass and ultrasonic sensor readings. If metric information is available, graph labels can include directions or distances to the neighbouring vertices. This allows not only for a wider spacing between snapshots but also to find shortcuts between snapshot chains over unknown terrain. A generalization of purely topological maps are graphs where edges are labelled by actions (e.g., Kuipers and Byun, 1991; Schölkopf and Mallot, 1995; Bachelder and Waxman, 1995). This way, systems can be built which do not depend on just one type of action (in our case this was a homing procedure). Although presented for navigation problems, similar graph approaches may well be feasible for other cognitive planning tasks, as, e.g., in the means-end-fields of Tolman (1932).

Clearly, the system we presented here is extremely simple compared to biological systems. Our intention is not to build models of animals, but to identify some of the basic building blocks that might play a role in biological navigation. This focus on understanding and synthesizing behaviour in a task-oriented way leads to parsimonious solutions with both technological and ethological implications.

Acknowledgments

The present work has profited from discussions and technical support by Philipp Georg, Susanne Huber, and Titus Neumann. We thank Fiona Newell and our reviewers for helpful comments on the manuscript. Financial support was provided by the Max-Planck-Gesellschaft and the Studienstiftung des deutschen Volkes.

Note

1. If the views are recorded using sensors with overlapping Gaussian receptive fields, the view will be a smooth function of the position.

References

Bachelder, I.A. and Waxman, A.M. 1995. A view-based neurocomputational system for relational map-making and navigation in visual environments. *Robotics and Autonomous Systems*, 16:267–289.

Brooks, R.A. 1986. A robust layered control system for a mobile robot. *IEEE Journal of Robotics and Automation*, RA-2(1).

Cartwright, B.A. and Collett, T.S. 1983. Landmark learning in bees. *J. Comp. Physiol. A*, 151:521–543.

Chahl, J.S. and Srinivasan, M.V. 1996. Visual computation of egomotion using an image interpolation technique. *Biol. Cybern.*, 74:405–411.

Collett, T.S. 1992. Landmark learning and guidance in insects. *Phil. Trans. R. Soc. Lond. B*, 337:295–303.

Collett, T.S. 1996. Insect navigation en route to the goal: Multiple strategies for the use of landmarks. *J. Exp. Biol.*, 199:227–235.

Franz, M.O., Schölkopf, B., and Bülthoff, H.H. 1997. Homing by parameterized scene matching. In *Fourth Europ. Conf. on Artificial Life*, P. Husbands and I. Harvey (Eds.), MIT Press: Cambridge, MA, pp. 236–245.

Franz, M.O., Schölkopf, B., Georg, P., Mallot, H.A., and Bülthoff, H.H. 1997. Learning view graphs for robot navigation. In *Proc. 1. Intl. Conf. on Autonomous Agents*, W.L. Johnson (Ed.), ACM Press: New York, pp. 138–147.

Gallistel, R. 1990. *The Organization of Learning*, MIT Press: Cambridge, MA.

Gillner, S. and Mallot, H.A. 1997. Navigation and acquisition of spatial knowledge in a virtual maze. *J. Cognitive Neuroscience*. In press.

Goldman, S. 1953. *Information Theory*, Dover: New York.

Hong, J., Tan, X., Pinette, B., Weiss, R., and Riseman, E.M. 1991. Image-based homing. In *Proc. IEEE Intl. Conf. on Robotics and Automation*, pp. 620–625.

Kuipers, B.J. and Byun, Y. 1991. A robot exploration and mapping strategy based on a semantic hierarchy of spatial representations. *Robotics and Autonomous Systems*, 8:47–63.

Lieblich, I. and Arbib, M.A. 1982. Multiple representations of space underlying behavior. *Behavioral and Brain Sciences*, 5:627–659.

Mallot, H., Bülthoff, H., Georg, P., Schölkopf, B., and Yasuhara, K. 1995. View-based cognitive map learning by an autonomous robot. In *Proc. ICANN'95—Int. Conf. on Artificial Neural Networks*, F. Fogelman-Soulié and P. Gallinari (Eds.) EC2, Nanterre, France, Vol. II, pp. 381–386.

Mataric, M.J. 1991. Navigating with a rat brain: A neurobiologically-inspired model for robot spatial representation. In *From Animals to Animats*, J.-A. Meyer and S.W. Wilson (Eds.), MIT Press: Cambridge, MA.

O'Keefe, J. and Nadel, L. 1978. *The Hippocampus as a Cognitive Map*, Clarendon Press: Oxford.
O'Neill, M.J. 1991. Evaluation of a conceptual model of architectural legibility. *Environment and Behavior*, 23:259–284.
Piaget, J. and Inhelder, B. 1967. *The Child's Conception of Space*, Norton: New York.
Poucet, B. 1993. Spatial cognitive maps in animals: New hypotheses on their structure and neural mechanisms. *Psychological Rev.*, 100:163–182.
Röfer, T. 1995. Controlling a robot with image-based homing. Center for Cognitive Sciences, Bremen, Report 3/95.
Schölkopf, B. and Mallot, H.A. 1995. View-based cognitive mapping and path planning. *Adaptive Behavior*, 3:311–348.
Thrun, S. 1995. Exploration in active learning. In *The Handbook of Brain Theory and Neural Networks*, M.A. Arbib (Ed.), MIT Press, pp. 381–384.
Tolman, E.C. 1932. *Purposive Behavior of Animals and Men*, Irvington: New York.
Wehner, R., Michel, B., and Antonsen, P. 1996. Visual navigation in insects: Coupling of egocentric and geocentric information. *J. Exp. Biol.*, 199:129–140.
Yagi, Y., Nishizawa, Y., and Yachida, M. 1995. Map-based navigation for a mobile robot with omnidirectional image sensor COPIS. *IEEE Trans. Robotics Automat.*, 11:634–648.

Matthias O. Franz graduated with a M.Sc. in Atmospheric Sciences from SUNY at Stony Brook, NY, in 1994, and with a Diplom in Physics from the Eberhard-Karls-Universität, Tübingen, Germany, in 1995. He is currently a Ph.D. student at the Max-Planck-Institut für biologische Kybernetik. His main research activities are in autonomous navigation, egomotion extraction from optic flow and natural image statistics.

Bernhard Schölkopf received an M.Sc. degree in Mathematics from the University of London in 1992. Two years later, he received the Diplom in Physics from the Eberhard-Karls-Universität, Tübingen, Germany. He recently finished his Ph.D. in Computer Science at the Technische Universität Berlin, with a thesis supervised by Heinrich Bülthoff and Vladimir Vapnik (AT&T Research). He has worked on machine pattern recognition for AT&T and Bell Laboratories. His scientific interests include machine learning and perception.

Hanspeter A. Mallot studied Biology and Mathematics at the University of Mainz where he also received his doctoral degree in 1986. He was a postdoctoral fellow at the Massachusetts Institute of Technology in 1986/87 and held research positions at Mainz University and the Ruhr-Universität-Bochum. In 1993, he joined the Max-Planck-Institut für biologische Kybernetik. In 1996/97, he was a fellow at the Institute of Advanced Study, Berlin. His research interests include the perception of shape and space in humans and machines, as well as neural network models of the cerebral cortex.

Heinrich H. Bülthoff is Director at the Max-Planck-Institute for Biological Cybernetics in Tübingen, Germany and Honorar Professor at the Eberhard-Karls-Universität Tübingen. He studied Biology in Tübingen and Berlin and received his doctoral degree in 1980. He spent 3 years as a visiting scientist at the Massachusetts Institute of Technology and joined in 1988 the faculty of the Department of Cognitive and Linguistic Sciences at Brown University. In 1993, he was elected as Scientific Member of the Max-Planck Society and currently directs at the Max-Planck-Institute in Tübingen a group of about 30 biologists, computer scientists, mathematicians, physicists and psychologists working on psychophysical and computational aspects of higher level visual processes. Dr. Bülthoff has published more than 50 articles in scholarly journals in the areas of Object and Face Recognition, Integration of Visual Modules, Sensori-Motor Integration, Autonomous Navigation and Artificial-Life. He has active collaborations with researchers at Brown University, Massachusetts Institute of Technology, NEC Research Institute, Oxford University, Tübingen University, Tel Aviv University, University of Minnesota, University of Western Ontario, University of Texas and the Weizmann Institute of Science.

GPSR Compliance

The European Union's (EU) General Product Safety Regulation (GPSR) is a set of rules that requires consumer products to be safe and our obligations to ensure this.

If you have any concerns about our products, you can contact us on ProductSafety@springernature.com

In case Publisher is established outside the EU, the EU authorized representative is:

Springer Nature Customer Service Center GmbH
Europaplatz 3
69115 Heidelberg, Germany

Batch number: 10607924

Printed by Printforce, the Netherlands